Lecture Notes in Computer Science   13167

More information about this subseries at https://link.springer.com/bookseries/7409

Shelly Sachdeva · Yutaka Watanobe ·
Subhash Bhalla (Eds.)

Big-Data-Analytics in Astronomy, Science, and Engineering

9th International Conference on Big Data Analytics, BDA 2021
Virtual Event, December 7–9, 2021
Proceedings

Springer

Editors
Shelly Sachdeva
National Institute of Technology
Delhi, India

Yutaka Watanobe
University of Aizu
Aizu-wakamatsu city, Fukushima, Japan

Subhash Bhalla ⓘ
University of Aizu
Aizu-wakamatsu city, Fukushima, Japan

ISSN 0302-9743 ISSN 1611-3349 (electronic)
Lecture Notes in Computer Science
ISBN 978-3-030-96599-0 ISBN 978-3-030-96600-3 (eBook)
https://doi.org/10.1007/978-3-030-96600-3

LNCS Sublibrary: SL3 – Information Systems and Applications, incl. Internet/Web, and HCI

This Springer imprint is published by the registered company Springer Nature Switzerland AG
The registered company address is: Gewerbestrasse 11, 6330 Cham, Switzerland

Preface

The volume of data managed by computer systems has greatly increased in recent times due to advances in networking technologies, storage systems, the adoption of mobile and cloud computing, and the wide deployment of sensors for data collection. As a result, there are five attributes of data that pose new challenges—volume, variety, velocity, veracity, and value. To make sense of emerging big data to support decision-making, the field of big data analytics has emerged as a key research and study area for industry and other organizations. Numerous applications of big data analytics are found in diverse fields such as e-commerce, finance, healthcare, education, e-governance, media and entertainment, security and surveillance, smart cities, telecommunications, agriculture, astronomy, and transportation.

Analysis of big data raises several challenges such as how to process extremely large volumes of data, process data in real-time, and deal with complex, uncertain, heterogeneous, and streaming data, which are often stored in multiple remote storage systems. To address these challenges, new big data analysis solutions must be devised by drawing expertise from several fields such as big data processing, data mining, database systems, statistics, machine learning, and artificial intelligence. There is also an important need to build data analysis systems for emerging applications, such as vehicular networks, social media analysis, and time-domain astronomy, and to facilitate the deployment of big data analysis techniques in artificial intelligence-related applications.

The ninth International Conference on Big Data Analytics (BDA) was held during December 7–9, 2021. It was held jointly in virtual conference mode at the University of Aizu, Japan, and in person at the National Institute of Technology, Delhi (NITD), India. The proceedings in this book include 15 peer-reviewed research papers and contributions by keynote speakers and invited speakers. This year's program covered a wide range of topics related to big data analytics on themes such as networking; social media; search; information extraction; image processing and analysis; spatial, text, mobile and graph data analysis; machine learning; and healthcare.

It is expected that the research papers, keynote speeches, and invited talks presented at the conference will encourage research on big data analytics and stimulate the development of innovative solutions and their adoption in the industry.

The conference received 60 submissions. The Program Committee (PC) consisted of researchers from both academia and industry from many different countries. Each submission was reviewed by at least two, and at most three, Program Committee members and was discussed by the PC chairs before a decision was made. Based on the above review process, the Program Committee selected 15 full papers. The overall acceptance rate was 25%.

We would like to extend our sincere thanks to the members of the Program Committee and external reviewers for their time, energy, and expertise in providing support to BDA 2021.

Additionally, we would like to thank all the authors who considered BDA 2021 as the forum to publish their research contributions. The Steering Committee and the

Organizing Committee deserve praise for the support they provided. A number of individuals contributed to the success of the conference. We thank H. V. Jagadish, Masaru Kitsuregawa, Sanjay Chawala, Divyakant Agrawal, and Sanjiva Prasad for their insightful suggestions. We also thank all the keynote speakers and invited speakers. We would like to thank the sponsoring organizations including the National Institute of Technology, Delhi (NITD); the Indian Institute of Information Technology, Delhi (IITD), India; the University of Aizu, Japan; and the University of Delhi, India, as they deserve praise for the support they provided.

The conference received valuable support from the University of Aizu and the NITD for hosting and organizing the conference. At the same time, thanks are also extended to the faculty, staff members, and student volunteers of the computer science departments at the University of Aizu and at NITD for their constant cooperation and support.

December 2021

Shelly Sachdeva
Yutaka Watanobe
Subhash Bhalla

Organization

BDA 2021 was organized by the University of Aizu, Japan, and the National Institute of Technology, Delhi, (NITD), India.

Patrons

Ajay Sharma   NIT Delhi, India
Toshiaki Miyazaki   University of Aizu, Japan

General Chairs

Sanjiva Prasad   IIT Delhi, India
Shelly Sachdeva   NIT Delhi, India

Steering Committee Chair

Subhash Bhalla   University of Aizu, Japan

Steering Committee

Srinath Srinivasa   IIIT Bangalore, India
Masaru Kitsuregawa   University of Tokyo, Japan
Vasudha Bhatnagar   University of Delhi, India
H. V. Jagadish   University of Michigan, USA
Divyakant Agrawal   University of California, Santa Barbara, USA
Arun Agarwal   University of Hyderabad, India
Subhash Bhalla   University of Aizu, Japan
Nadia Berthouze   University College London, UK
Cyrus Shahabi   University of Southern California, USA

Program Committee Chairs

Yutaka Watanobe   University of Aizu, Japan
Shelly Sachdeva   NIT Delhi, India

Organizing Chair

Shelly Sachdeva   NIT Delhi, India

Publication Chair

Subhash Bhalla   University of Aizu, Japan

Convenor

Shivani Batra   KIET Group of Institutions, India

Tutorial Chair

Punam Bedi   Delhi University, India

Publicity Chair

Rashmi Prabhakar Sarode   University of Aizu, Japan

Panel Chair

Wanming Chu   University of Aizu, Japan

Website Chair

Manoj Poudel   University of Aizu, Japan

Website Administrator

Divij Gurmeet Singh   University of Aizu, Japan

Program Committee

D. Agrawal   University of California, USA
F. Andres   National Institute of Informatics, Japan
Nadia Bianchi-Berthouze   University College London, UK
Paolo Bottoni   University of Rome, Italy
Pratul Dublish   Microsoft Research, USA
William I. Grosky   University of Michigan-Dearborn, USA
Jens Herder   University of Applied Sciences Düsseldorf,
                                  Germany
Masahito Hirakawa   Shimane University, Japan
Qun Jin   Waseda University, Japan

Akhil Kumar	Pennsylvania State University, USA
Srinath Srinavasa	IIIT Bangalore, India
Rishav Singh	NIT Delhi, India
Jianhua Ma	Hosei University, Japan
Anurag Singh	NIT Delhi, India
K. Myszkowski	Max-Planck-Institut für Informatik, Germany
T. Nishida	Kyoto University, Japan
Alexander Pasko	Bournemouth University, UK
Manisha Bharti	NIT Delhi, India
Baljit Kaur	NIT Delhi, India
Prakash Srivastava	KIET Group of Institutions, India
Rahul Katarya	Delhi Technological University, India
Vivek Shrivastava	NIT Delhi, India

Sponsoring Institutions

National Institute of Technology, Delhi, India
Indian Institute of Technology, Delhi, India
University of Aizu, Japan

Contents

Data Science: Systems

Big Data Management for Policy Support in Sustainable Development

Pooja Bassin(ID), Niharika Sri Parasa(ID), Srinath Srinivasa(✉)(ID),
and Sridhar Mandyam(ID)

International Institute of Information Technology, 26/C, Electronics City Phase 1,
Bangalore, Karnataka, India
{pooja.bassin,niharikasri.parasa,sri,sridhar.mandyamk}@iiitb.ac.in

Abstract. Achievement of targets and indicators for Sustainable Development Goals (SDGs) are of key interest to policy-makers worldwide. Sustainable development requires a holistic perspective of several dimensions of human society and addressing them together. With the increasing proliferation of Big Data, Machine Learning, and Artificial Intelligence, there is increasing interest in designing Policy Support Systems (PSS) for supporting policy formulation and decision-making. This paper formulates an architecture for a PSS, based on combining data from several sources. These datasets are subject to cleaning and semantic resolution and are then used as inputs to support the building of semantic models based on Bayesian Networks. A set of models built for different SDG sub-goals and indicators is used to create a "Policy Enunciator" in the form of data stories. Policy formulation is supported by modeling interventions and counter-factual reasoning on the models and assessing their impact on the data stories. Two kinds of impacts are observed: (a) *downstream* impacts that track expected outcomes from a given intervention, and (b) *lateral* impacts, that provide insights into possible side-effects of any given policy intervention.

Keywords: Policy support system · Big data · Knowledge graph · Bayesian network

1 Introduction

Sustainable development as an overarching concern is increasingly adopted by several countries worldwide. Sustainable development as an organizing principle strives for economic progress in a way that nourishes and sustains natural mechanisms that human societies depend on. Sustainable development does not refer to only one sector of human activity and requires a holistic approach towards human development. A global agenda for sustainable development has been spearheaded by the United Nations (UN) in the form of 17 Sustainable Development Goals (SDGs)[1]. A set of targets and indicators have also been

---

[1] UN Sustainable Development Goals https://sdgs.un.org/.

© Springer Nature Switzerland AG 2022
S. Sachdeva et al. (Eds.): BDA 2021, LNCS 13167, pp. 3–15, 2022.
https://doi.org/10.1007/978-3-030-96600-3_1

developed as a global agenda for sustainable development by 2030[2]. The UN SDGs are a set of 17 broad goals that define a call for action for nations across the world.

Each of the 17 goals are further divided into several specific targets and indicators that provide actionable and measurable insights about the progress towards these goals. For instance, the first target under SDG 1 (No poverty), is as follows: "By 2030, eradicate extreme poverty for all people everywhere, currently measured as people living on less than US$1.25 a day."

SDGs are declarative and do not by themselves mandate a specific policy or procedure to achieve these targets. It is also widely acknowledged by nations that when it comes to policies and practices, no one size fits all. To achieve the same target, each nation would need to develop customized policies and practices relevant to its context, and also break down the goals into sub-national and local levels[3].

With increasing digitization in administration and social practices, there is an increasing interest in adopting data-driven platforms and practices for the complex endeavour of re-purposing SDGs at sub-national and local levels [10, 16, 17]. While Decision Support Systems (DSS) and various kinds of data warehousing solutions have been around for a long while, they are inadequate in themselves, to address the complexities of policy support. While different policies are handled by disparate departments, there are a wide variety of cross-linkages across policies and their instruments across domains, some of which may change over time. For instance, a new policy on mining in forest areas impacts the environment and ecology directly, but has many indirect impacts on livelihoods of people living in those neighbourhoods, land and water utilization, prices of commodities, rural incomes from ancillary units, and many more. Legacy systems do not provide the capability to uncover and handle such complex inter-linkages, nor the analytical tools to model them.

In this work, we explore the task of designing policy support systems for SDGs by managing data from disparate data sources. A characteristic element of the proposed platform is the policy enunciator, which, unlike analytics performed within an overarching schematic structure, supports dynamic discovery and construction of semantic models, with repercussions of policy interventions being tracked across multiple models.

As a prototype implementation, we consider one of the SDGs (SDG 2: Zero Hunger) and characterize its targets and indicators. A collection of disparate but pertinent data sets is then used to build a knowledge graph, and a set of semantic models to explore policy formulation and intervention mechanisms.

---

[2] Transforming our world: the 2030 Agenda for Sustainable Development https://sdgs. un.org/2030agenda.

[3] No 'one size fits all' approach for localisation of SDGs: India at UN https://www. hindustantimes.com/india-news/no-one-size-fits-all-approach-for-localisation-of-sdgs-india-at-un-101625802791716.html.

2 Related Literature

Decision support systems (DSS) have been around since the 1960s and have been widely deployed for supporting strategic decision-making. Power et al. [14] provides a categorization of DSS that includes: communications-driven, data-driven, document-driven, knowledge-driven, and model-driven systems. Each belonging to a specific category of input and skill set, aim to achieve a common goal of decision-making. However, a common characteristic of a DSS is to support decision-making for specific questions asked within a larger organizational framework. The prevalence of Big Data coming from several sources has ushered in a management revolution that allows businesses to make better predictions and smarter decisions using data-driven models [11]. Vydra and Klievink [22] argue that big data in terms of volume, velocity, variety, veracity, and sometimes variability, value, and visualization are the set of attributes that describe the data in a techno-optimistic way. However, the social change motivated by big data is policy-pessimist due to slow political and bureaucratic decision-making in contrast with rapidly changing technology. As a result, there has been increasing interest in designing Policy Support Systems (PSS), which are technologies designed for policy formulation and the intervention itself, rather than just supporting specific decisions. Guppy [9], proposes a Big Data-based policy support environment called SDG-PSS[4], to effectively deliver on SDG 6 (clean water and sanitation) targets[5]. Steps formulated on the components of SDG-6 primarily focus on the capability to implement and ensure that the implementation leads demonstrably to sustainability. This will ensure countries see these components as part of their enabling environment. A web-based spatial policy support system has been initiated for the development and implementation of conservation strategies focused on sustaining and improving ecosystem services[6]. The application of this tool can be found in [12] to identify priority areas for sustainable management to realize targets under SDG 6 at the country scale for Madagascar and the basin-scale for the Volta Basin. Within this SDG 6 priority areas footprint, they assess the synergies and trade-offs provided by this land for SDG 15 (life on land/biodiversity) and SDG 13 (climate action) as well as SDG 2 (zero hunger). The need for a PSS has been expressed in different ways by several other researchers as well. Griggs et al. [7] stress the need for a systemic approach to achieving SDGs and the need to go beyond decision-making in silos. They also propose an integrated framework [6] combining six SDGs to collectively track indicators and targets, as well as track the ways interventions towards one target affect others. Thomas et al. [19] emphasize the need for adaptive and dynamic models to support policy making in the context of SDGs and how interventions are often rendered ineffective due to external factors. Unlike a DSS, Policy Support System is not limited to strategic decision making, rather takes into consideration an all-round approach to investigate the robustness of a policy.

---

[4] SDG 6 Policy Support System: https://sdgpss.net/en/.

[5] UN SDG Goal 6: https://sdgs.un.org/goals/goal6.

[6] CostingNature: http://www.policysupport.org/costingnature.

With its minimal scope, DSS snubs cross-disciplinary attitudes but focuses predominantly on the local context of the theme. The notion behind PSS is to develop such a procedure that encompasses multiple heterogeneous but related data sources at one location, study the interactions among them and assess the viability of policy options spanning across numerous frameworks within the support system.

3 Approach

In this paper, we propose the architecture of a Policy Support System named *Poura* (meaning "citizen"), to assist policy researchers in the design and evaluation of policies with a particular focus on SDGs. The task of formulating, implementing, and evaluating public policies to achieve SDGs fall largely under the ambit of governments through governmental and non-governmental institutions. Even so, private institutions and individuals have significant roles to play towards ensuring that the target actors conform to and comply with public policy instruments to achieve stated sustainable goals on socio-economic and environmental objectives.

SDG policy-makers today are faced with new challenges. On the one hand, they need to use data and evidence-based reasoning for assessing the impact of policy instruments. On the other, they are faced with increasing uncertainties in both the economic and environmental world, on which they are attempting to make predictions. There is also a new recognition of the need to formulate policies that can handle shocks from "black-swan" events [1,2], making the socio-economic milieu resilient and robust. At the same time, there is a desire to endow economies and social systems with the flexibility to adapt to complex changes that may not be foreseen accurately.

The proposed PSS comprises three overall components supporting a policy researcher, which are discussed in the architecture in Fig. 1.

The three components of the Poura PSS are as follows:

3.1 Data Acquisition System (DAS)

This component is meant to process datasets and perform operations like data cleaning, semantic resolution, canonicalization, etc. A set of scripts ingest a wide variety of data such as structured, semi-structured, unstructured, or binary data into a data lake from disparate open data sources. It is equipped with a flat architecture, where every data element in the lake is given a unique identifier and tagged with a set of metadata information. The DAS comprises three operational layers, which are described as follows.

The *ingestion layer* acts a storage layer for raw data that is incoming to system. Data is ingested via connectors from various open data sources like the following: Open Government Data (OGD) Platform India[7], e-National

---

[7] OGD: https://data.gov.in.

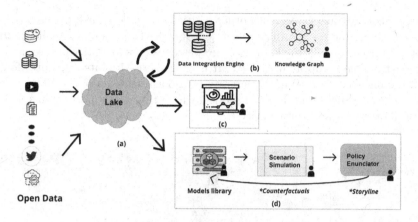

Fig. 1. Poura architecture

Agriculture Market[8], Central Control Room for Air Quality Management[9], and United Nations Statistics Portal[10]. Multiple approaches of ingestion are supported including batch uploads, real-time subscription, and one-time loading of datasets. The layer also offers an option to apply schema or meta-data information to the incoming data.

The *caching layer* is to temporarily or permanently store processed or preprocessed data and materialized views. The data in this layer is either ready for visualization and consumption by external systems or is prepared for further processing. Applications residing in the processing layer will take data from the ingestion layer, process it, structure it, and store them back in the data lake.

The *processing layer* or consumption layer is to offer one or more platforms for distributed processing and analysis of large data sets. It can access data stored in both the ingestion and caching layers. This pre-processed data is pushed to the latter stages to perform data driven, model driven, and knowledge driven analysis.

One of the main functionalities of the DAS is to curate the data in a way that is accessible for further downstream operations. Incoming data are of different varieties including tabular data, natural language text, social media posts, tweets, audio recordings, videos, etc. First, all multi-media datasets are converted to text, with semi-automatic transcription using available tools. Once all forms of data are converted into texts, they are further cleaned and represented in one of two different forms: a collection of tables, and a knowledge graph.

A set of scripts for semantic resolution are also used to identify semantically meaningful entities from the data, by resolving them with ontologies from the Linked Open Data (LOD) cloud[11]. An application-wide knowledge graph is also

---

[8] ENAM: https://enam.gov.in/.
[9] NAQI: https://app.cpcbccr.com/AQI_India/.
[10] UN statistics portal: http://data.un.org/.
[11] https://lod-cloud.net/.

built, by connecting entities with other entities through labeled, directed edges, and characterizing each entity with its attributes. This process is called *entity twinning*. Each node in the knowledge graph represents an entity of interest like a district, village, crop, industry, etc. It is characterised by its attributes found from the ingested data. With each entity is also associated a set of tables where this entity may be found, a set of models where this entity participates, and a set of data stories about the entity.

Fig. 2. Poura knowledge graph node

Figure 2 shows an example node from the knowledge graph rendered as a web page. Information about the node is divided into four sections: About (containing attributes and description obtained from the ingested data), Models, Datasets, and Data Stories.

3.2 Policy Analytics System

The next subsystem for Poura is the Policy Analytics Subsystem (PAS). This subsystem inter-operates with the DAS both as a downstream component of analytical processing, as well as an upstream component, where latent relationships discovered by this subsystem go on to curate the tables and knowledge graph in the DAS.

PAS has two broad classes of capabilities in the form of models libraries. These are described as follows:

Discovery-Driven Analytical Models Library: Unravelling of dependencies and or provable causalities between data elements is an essential perquisite for gaining an understanding of the cross-linkages between measures related to SDGs. This is typically a task of learning about conditional dependencies and structures in diverse data, often modelled through Directed Acyclic Graphs (DAGs) in a Bayesian probabilistic setting. Many traditional AI/ML techniques, aside from advanced statistics also assist in gaining an understanding of latent relationships in data. These models may be seen as forming one layer of a discovery-driven network connecting diverse data elements as nodes to one another through conditional dependencies, or perhaps even causality.

Prescriptive Models Library: This library comprises models that are built from domain knowledge with expert oversight. Many policy research questions relating to SDG goals straddle social and economic concerns, which have a vast pedagogy of prior research where established principles of micro/macro-economics relating to markets, labour, finance, pricing, demand, etc. all of which play a role in the achievement of SDGs. Many well-understood concepts have empirical roots and a plethora of mathematical models that capture the essence of socio-economic linkages. The PSS system is required to be designed in a manner that a set of such normative models, borrowed from other economic and social sciences, may be plugged into work on the data acquired.

This set of models thus may be viewed as another network layer interconnecting the set of data elements, in addition to the knowledge graph. It is also conceivable that the Bayesian learning models which uncover latent conditional dependencies in the first layer above, can inform the development and tuning of normative models in the second layer.

Figure 3 shows a Bayesian network model depicting how various factors affect the average food supply in a region. This formal model is built from studying factors affecting food supply documented by the UN Food and Agriculture Organization (FAO), FAOSTAT[12]–the statistical division of FAO. FAOSTAT provides free access to food and agriculture data for over 245 countries for roughly 60 years. The linkages between the nodes in the DAG attempt to establish causal dependencies. The nodes represent elements from significant agriculture domains such as food production, land use, climate change, pesticides, etc. The target variable, the average food supply network, seeks to identify the food supplies available for human consumption in caloric value.

A Bayesian network model [13] is in the form of a Directed Acyclic Graph (DAG) with a single "target" node, representing the variable as the desired outcome. Arcs connecting nodes represent conditional probabilities by which the antecedent node affects the consequent. The network structure represents

---

[12] FAOSTAT: https://www.fao.org/faostat/.

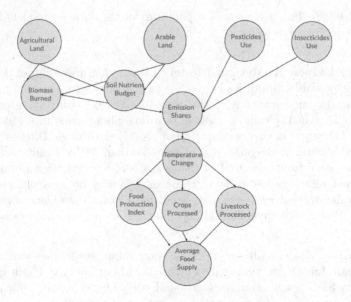

Fig. 3. FAO average food supply model

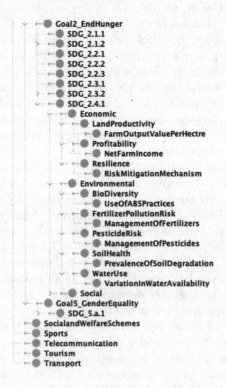

Fig. 4. Ontology describing SDG goals, subgoals and indicators

how the joint probability of all the variables factorizes based on conditional dependencies.

In addition to the Bayesian Network models, the PAS also hosts an ontology describing the SDGs, their subgoals, targets, and indicators. Figure 4 shows a snapshot of the ontological structure depicting SDG goals, subgoals, targets, and indicators. Bayesian Network models are aligned to different parts representing targets in this ontology-based on their target variable.

3.3   Policy Enunciator System

The third subsystem of the PSS is the *Policy Enunciator* subsystem (PES). This subsystem is meant for interaction with the end-user who is concerned with policy formulation. The primary design element of the PES is the "language" in which policy related issues are enunciated.

PES represents policy related issues using a combination of two elements: *model ensemble* and *data story.*

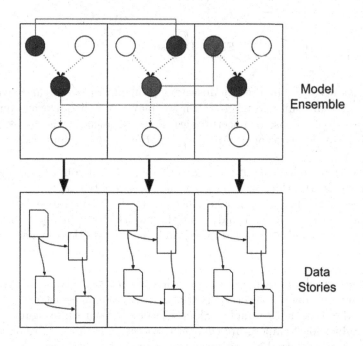

Fig. 5. Policy enunciator: models ensemble and data stories

Figure 5 schematically depicts the policy enunciator components. The PES comprises of a *models ensemble* which is a set of models from the models library that have one or more variables in common. These variables act as confounding variables across different concerns. In the figure, confounding variables are shown in colour, and are coupled with their counterpart in another model.

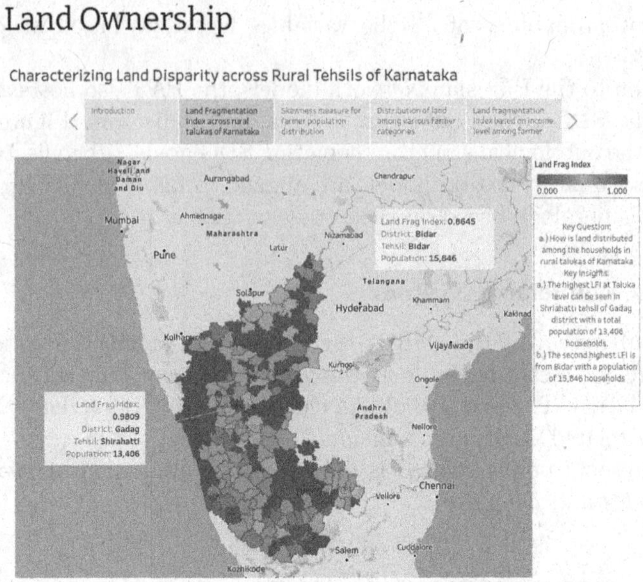

Land Ownership

Characterizing Land Disparity across Rural Tehsils of Karnataka

Fig. 6. Data story

Bayesian networks are used for answering probabilistic queries and in a policy scenario, they are suitable where when we may ask policy-relevant questions to evaluate them in case of uncertainties. Policy enunciation is carried out by performing two kinds of operations: *intervention modeling* and *counterfactual analysis*.

Intervention modeling starts by converting a proposed policy instrument into one or more interventions in a Bayesian model, to set the value of variables to specific levels. For instance, a policy instrument that strives to reduce the use of pesticides in agriculture would affect the model shown in Fig. 3, by setting the values of the nodes "Pesticide Use" and "Insecticide Use" to low levels. In Bayesian reasoning, such manual intervention into a model is constructed using the **do()** operation [5].

Counterfactual analysis, also called "what-if" analysis also performs similar interventions on models to set values to variables that may not necessarily be visible in the data. For example, taking the model from Fig. 3 again, suppose that the values for "temperature change" recorded in the data were only minor changes. The model would have learned conditional probabilities of its effect on other downstream variables based on the data available. This model can then be used to set "temperature change" to a higher value, to see its impact on downstream variables.

The impact of interventions may lead to two kinds of repercussions: a *downstream* impact where a change in one variable may have a chain of events within a given model, and/or a *lateral* impact, where a change in variable impacts some other model(s) due to its confounding role in the other model(s). For instance, a decision to impose a lockdown in the face of a raging pandemic would have a downstream impact on curbing the spread of the infection. But it would also have several lateral impacts affecting other aspects of human life like the economy, mental health, etc. The more and richer the models we have for different aspects of governance, the better we will be able to track lateral repercussions.

Policy enunciation to track both intended and collateral consequences is perhaps the primary motivation for building Policy Support Systems.

The probability computations estimated by the network along with the statistical history of entities presented as *data stories* help in assimilating the implication of data. The objective is to enhance the engagement of policy makers with the PSS towards capturing total policy perspective to make wise policy decisions. Figure 6 refers to landownership story for rural tehsils of Karnataka. Based on available data, the story attempts to address some key questions related to ownership of small or large piece of land being determinant to understand the veracity of economic status of households in rural Karnataka. A data story is in the form of a sequence or "storyline" of analytic results, that builds a narrative flow concerning any policy matter of interest. For instance, the data story presented in Fig. 6 asks whether land fragmentation is an indicator of economic status of households in the tehsils of Karnataka, and to understand how policy changes impacting land fragmentation would affect economic status of households.

A data story is specified by a template, which is populated from data available in the data lake as well as relevant model outputs. Intervention and counterfactual analysis resulting in changes to one or more models, result in corresponding changes in data stories. A data story is also associated with a theme in the SDG ontology. Thus every proposed intervention can be tracked to different SDG themes that it may likely affect, and detailed using the corresponding data stories.

4 Conclusions and Future Work

Decision Support Systems have been developed in various domains including clinical [15,20], agricultural [3,18,21], logistics [4], forecasting [8] and many other areas with a specific structure designed to cater and support definite set of problems. The Policy Support System intends to aid policymakers in understanding different spheres of sustainable development goals' targets and indicators. The SDG targets have common concerns across domains hence undeniably these elements play a significant role affecting multiple terrains. The distinguishing feature of a PSS is the ability to identify intervention impacts within and beyond models featuring SDG related themes. Data stories communicate insights using visualisations that facilitate in directing attention to such impacts for policy interventions. These in turn are valuable in achieving the goals in a timely manner.

The models library presently comprises of elements related to SDG themes that may be expanded using expert knowledge at a more granular level. We plan to expand the knowledge graph with all possible entities pertaining to governance of Karnataka districts and identify latent relations to strengthen discovery driven analytics. Building a *District-SDG* tracking map that plots all the districts of Karnataka with their current SDG ranks as origin points shall allow researchers and policymakers to understand and comparatively analyze the policy decisions in a regional context to achieve target indicators. This may also assist in tracking the progress of a district vis-á-vis an SDG goal by intervening with the model attributes and utilizing formerly derived data. This would lead to identify and understand the attributes impacting progress in a district and which of them to intervene to reach a target goal.

References

1. Aven, T.: On the meaning of a black swan in a risk context. Saf. Sci. **57**, 44–51 (2013)
2. Batrouni, M., Bertaux, A., Nicolle, C.: Scenario analysis, from bigdata to black swan. Comput. Sci. Rev. **28**, 131–139 (2018)
3. Gandhi, N., Armstrong, L.J., Petkar, O.: Proposed decision support system (DSS) for Indian rice crop yield prediction. In: 2016 IEEE Technological Innovations in ICT for Agriculture and Rural Development (TIAR), pp. 13–18. IEEE (2016)
4. George, P., Subramoniam, S., Krishnankutty, K.: Web based real-time DSS for cement supply chain and logistics. In: 2012 International Conference on Green Technologies (ICGT), pp. 001–005. IEEE (2012)
5. Gillies, D.: Causality: models, reasoning, and inference judea pearl. Br. J. Philos. Sci. **52**(3), 613–622 (2001)
6. Griggs, D., et al.: An integrated framework for sustainable development goals. Ecol. Soc. **19**(4), 49 (2014)
7. Griggs, D., et al.: Sustainable development goals for people and planet. Nature **495**(7441), 305–307 (2013)
8. Kiresur, V., Rao, D.R., Sastry, R.K., et al.: Decision support systems [DSS] in forecasting of future oilseeds scenario in India-a system dyanamic model. In: Proceedings of the First National Conference on Agri-Informatics (NCAI), Dharwad, India, 3–4 June 2001, pp. 90–93. Indian Society of Agricultural Information Technology (INSAIT) (2002)
9. Guppy, L.: Accelerating water-related SDG success; 6 steps and 6 components for SDG 6. UNU-INWEH Policy Brief, Issue 4 United Nations University Institute for Water Environment and Health. Hamilton, Ontario, Canada (2017)
10. Malhotra, C., Anand, R., Singh, S.: Applying big data analytics in governance to achieve sustainable development goals (SDGs) in India. In: Munshi, U.M., Verma, N. (eds.) Data Science Landscape. SBD, vol. 38, pp. 273–291. Springer, Singapore (2018). https://doi.org/10.1007/978-981-10-7515-5_19
11. McAfee, A., Brynjolfsson, E., Davenport, T.H., Patil, D., Barton, D.: Big data: the management revolution. Harv. Bus. Rev. **90**(10), 60–68 (2012)
12. Mulligan, M., van Soesbergen, A., Hole, D.G., Brooks, T.M., Burke, S., Hutton, J.: Mapping nature's contribution to SDG 6 and implications for other SDGs at policy relevant scales. Remote Sens. Environ. **239**, 111671 (2020)
13. Pearl, J.: Bayesian networks: a model CF self-activated memory for evidential reasoning. In: Proceedings of the 7th Conference of the Cognitive Science Society, University of California, Irvine, CA, USA, pp. 15–17 (1985)

14. Power, D.J.: Decision support systems: a historical overview. In: Burstein, F., Holsapple, C.W. (eds.) Handbook on Decision Support Systems, vol. 1, pp. 121–140. Springer, Heidelberg (2008). https://doi.org/10.1007/978-3-540-48713-5

15. Praveen, D., et al.: Smarthealth India: development and field evaluation of a mobile clinical decision support system for cardiovascular diseases in rural India. JMIR Mhealth Uhealth **2**(4), e3568 (2014)

16. del Río Castro, G., Fernández, M.C.G., Colsa, Á.U.: Unleashing the convergence amid digitalization and sustainability towards pursuing the sustainable development goals (SDGs): a holistic review. J. Cleaner Prod. **280**, 122204 (2020)

17. Sarker, M.N.I., Wu, M., Chanthamith, B., Yusufzada, S., Li, D., Zhang, J.: Big data driven smart agriculture: pathway for sustainable development. In: 2019 2nd International Conference on Artificial Intelligence and Big Data (ICAIBD), pp. 60–65. IEEE (2019)

18. Shirsath, R., Khadke, N., More, D., Patil, P., Patil, H.: Agriculture decision support system using data mining. In: 2017 International Conference on Intelligent Computing and Control (I2C2), pp. 1–5. IEEE (2017)

19. Thomas, A., et al.: Rapid adaptive modelling for policy support towards achieving sustainable development goals: brexit and the livestock sector in wales. Environ. Sci. Policy **125**, 21–31 (2021)

20. Usmanova, G., et al.: The role of digital clinical decision support tool in improving quality of intrapartum and postpartum care: experiences from two states of India. BMC Pregnancy Childbirth **21**(1), 1–12 (2021)

21. Vishwajith, K., Sahu, P., Dhekale, B., Mishra, P., Fatih, C.: Decision support system (DSS) on pulses in India. Legume Res. Int. J. **43**(4), 530–538 (2020)

22. Vydra, S., Klievink, B.: Techno-optimism and policy-pessimism in the public sector big data debate. Gov. Inf. Q. **36**(4), 101383 (2019)

Development of the Research Data Archive
at OIST

Matthew R. Leyden(✉) (iD), Nicholas M. Luscombe(iD), and Milind Purohit

Okinawa Institute of Science and Technology Graduate University, 1919-1 Tancha, Onna-son,
Kunigami-gun, Okinawa 904-0495, Japan
Matthew.Leyden@OIST.jp

Abstract. Once a work is published academic authors may not spend much time
thinking about its data any longer. If an author decides to revisit a similar subject,
or share old data with a colleague, having properly formatted original data can
make that easier. Unfortunately, files and documentation may get lost or misplaced
as students and postdocs change jobs, or file types become obsolete and difficult
to access. Research data is valuable to authors, universities, and the scientific
community. Many funding institutions around the world are now requiring data
management plans to be submitted along with a research proposal [1]. However,
it can be difficult for funders to evaluate if data plans are properly exercised.
The Okinawa Institute of Science and Technology (OIST) seeks to be among the
first institutions that have a centralized copy of all its published research data. As
of this fiscal year, OIST now requires the curated archival of published research
data. This ensures that not only the final published document is available, but
also the raw data and information necessary to reach the conclusions presented
in the publication. OIST strives to make sure that this data stays organized and
easily accessible in the years to come. However, archiving requires a high level
of cooperation from authors, and it cannot be a one-sided effort by the university
administration. This conference proceeding covers our motivations, experiences,
challenges, and summarizes the preliminary results from our first 6 months of
archiving.

Keywords: Archive · Publication data · Data reuse

1 Introduction

Archiving data has many motivations, but primarily it is to preserve data for future use.
One might assume that authors are careful to save a copy of their data. However, one
study analyzing data in biology found that research data disappears at a rate of 17%
per year [2]. In another example, NASA lost a video of the Apollo 11 moon landing
because tapes were overwritten [3]. In short, data loss is common, and it can happen on
high-profile projects. Even if data is not completely lost, it will be increasingly difficult
to find and use data as time goes on if it is not properly archived. Some researchers may
leave, and others may begin to forget details about how the data was collected. If data

S. Sachdeva et al. (Eds.): BDA 2021, LNCS 13167, pp. 16–25, 2022.
https://doi.org/10.1007/978-3-030-96600-3_2

is properly organized, documented, and saved in a secure location, data stands a much higher chance of being useful for future research.

Another motivation is to share data with the research community, this allows for data reuse in future studies [4]. The publication by itself is useful for communicating key concepts, but for another researcher to create their own conclusions they will likely need direct access to original data. Archiving can facilitate this by directly sharing research data, or simply by keeping data ready and organized to be shared if desired by the authors. Research data is often very expensive to generate, so it is a more efficient application of research funds if authors can reuse data.

A third motivation is to prevent scientific fraud. Fraud does not happen often but can lead to significant consequences. One notable example was the case of STAP cells [5], resulting in the suicide of one of the co-authors and an element of notoriety for the university (RIKEN). As a consequence of this event and others, in 2014 the Ministry of Education, Culture, Sports, Science and Technology (MEXT) created Guidelines for Responding to Misconduct in Research [6]. These guidelines instruct institutions to adopt rules requiring researchers to keep research data.

In response to the motivations listed above, in 2017 OIST updated its rules and procedures mandating archiving and provided guidelines for researchers. In 2021, OIST began to actively archive data, with efforts led by a data archive coordinator. To make this feasible, OIST limited the scope to only published data. Unpublished data is constantly changing, poorly defined, and likely undocumented. Consequently, a broad policy on the archival of unpublished data will likely not be implemented soon. However, we aim for the proper storage and archival of data to be increasingly inclusive. OIST has centralized data storage, and a portion of that is devoted to the archive. Diagrams summarizing the archival process are shown in Fig. 1. Data is submitted to the coordinator, who checks if the submission complies with the standards set by OIST. This submission is essentially private to the research unit and is not shared beyond those with appropriate permissions. OIST does not mandate sharing data, which allows researchers to maintain some privacy. If authors were forced to share, they may decide to omit important information. If authors are confident that their data is safe, they may be more willing to submit a complete data set to the archive. For similar reasons, the archive is currently only accepting fully anonymous human subject data.

OIST tries to set a uniform standard for archival for all its researchers. Setting standards comes with some challenges as university research is heterogeneous. For this reason, the standards are slightly abstract. We ask that our authors include a copy of the published data in its original format and a copy of the final data presented in the manuscript. Data should include code and analysis that was used to process the original data into its final state. Also, we ask authors to provide a satisfactory level of documentation, such that a motivated researcher should be able to trace from the original measurement to the conclusions of the paper. This documentation should include important information about the data, such as, who created it, when was it taken, on what system was it measured, and how unusual file types can be accessed. Some of these details are often expressed in the manuscript, and in general, we try to avoid exasperating authors by giving them redundant work. However, it is not uncommon that specifics are vague or omitted from the publication. The coordinator is motivated to resolve details that might get overlooked during the publication review. In aggregate, we set these standards to ensure that the work is reproducible, and any allegations of fraud can be easily investigated.

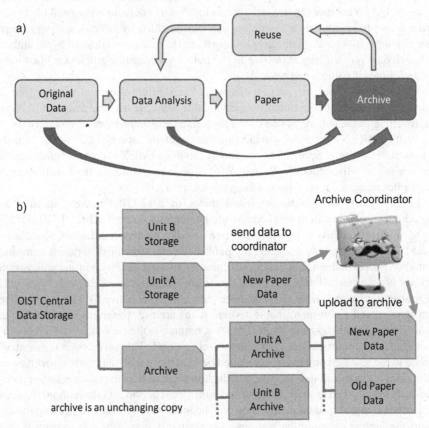

Fig. 1. Overview of the archival process. a) The diagram illustrates the flow of information during archival, highlighting the reuse of data for future publications. b) Storage diagram for the OIST publication archive. Research units all have their private storage and archive that are essentially only visible to them. The archive is a read-only copy curated by a coordinator.

2 Results and Discussion

Before implementing policy, the coordinator tried to meet with faculty to discuss their data in the context of archiving. This meeting ensured that all faculty were aware of the university's efforts, and helped the coordinator better understand the needs of the research units. We were able to survey 71 of the total 81 faculty. The results are summarized below in Fig. 2. Every OIST unit is given at least 50 TB of storage, and 35% of faculty mentioned that they were near their storage capacity. We did not plan to allot more research space as part of the archive initiative but can alleviate some units by allowing them to transfer some of their old data to the archive. We asked faculty to estimate the size of data they would expect from a typical paper. In these meetings we noticed that faculty occasionally expressed resistance to the idea of archiving. We have summarized the prevalence of some common concerns in Fig. 2. The most common concern was that archiving would create additional workload. There were also concerns about privacy, and data inspection by a coordinator.

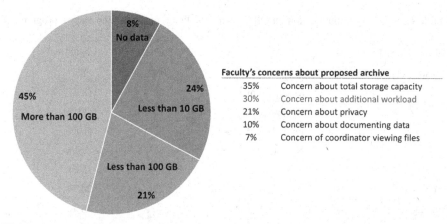

Fig. 2. Survey results of faculty on data archiving. The survey indicated that 100 GB per publication will likely be enough for most faculty. This table summarizes some of the common concerns expressed by faculty, such as total space, and data privacy. The most significant concern affecting the archive was the additional workload from archiving requirements.

As of April 2021, OIST began actively seeking the archival of published research. To limit costs and ensure that the data is a manageable size, we decided to limit the archive submission size to 100 GB per publication. Our survey suggested that 54% of faculty would likely have their needs met by 100 GB of storage per paper. This will cover most papers but admittedly does not address the data storage needs of many. We currently ask authors with publication data beyond the 100 GB cap to choose the most important data to store in the archive. Other data should be kept in a secure location and documentation should note where this data is stored. We chose to prioritize keeping things simple for the first year of archiving to facilitate a successful start. In the years to come, we may decide to increase the cap if there is an obvious need, and the budget allows. Later we will show that the average submission to the archive is well within this limit.

Even though OIST only recently implemented its archiving program, we still can share our preliminary results. Hopefully, this data can be used to estimate how a similar initiative may perform at another university. Table 1 shows a summary of OIST's publications during the first 6 months of archiving. In this time, the coordinator recorded 279 publications, and 79 of these have been archived. Most papers are not subject to our archive requirements. This may be surprising at first, but there are many reasons why a paper may not submit to the archive. Often a publication was an effort of multiple universities, where OIST was not directly responsible for the data (~29% of publications). There were many instances when there was no data to archive (~19% of publications), such as a review paper, or a purely analytical paper. OIST does not require the archival of data that is already publicly available, so long as the data offered is reasonably consistent with the standards of OIST. We do not wish to discourage public sharing and avoid asking authors to submit their data multiple times. Because we have just started archiving a significant fraction of requests for data are still pending. Pending requests will likely become less significant with time.

Table 1. Summary of publications at OIST during the first 6 months of the archive program.

Publication status		Count	% of total
Not subject to archive		165	58%
	Multiple universities	(82)	(29%)
	Review paper	(30)	(11%)
	Publicly available	(27)	(10%)
	No data	(23)	(8%)
	Other	(3)	(1%)
Archived		**79**	**28%**
Pending resolution		40	14%
Total papers		284	

This is obvious, but the first step in publication data archival is to be aware a publication exists. We wish to get the data soon after a paper is accepted, so data can be collected before authors leave, or begin to forget about important details. We ask authors to self-report their publications to the coordinator. However, we found that less than 5% of publications were self-reported before the coordinator was able to find them via a search engine. Therefore, we can advise that archival will likely be resolved more promptly and completely if publications are recorded by a motivated party such as the university library or an archive coordinator. At OIST, the coordinator keeps a record of recent publications, and once aware of a publication, sends out an email request to relevant authors for data. In June, we began tracking request dates and resolution dates. This data is shown in Fig. 3. If it was obvious that the paper contains no appropriate data, no request was sent. If authors clarify why they have no data to archive, this is included in the figure. If the authors did not respond after 1 month, another reminder email was sent. Authors that have yet to resolve our request for data are included, where the response time is calculated from the date of preparing this figure (2021/11/12). Under these conditions, we found that the average response time was 28 days.

After 79 submissions, we can make some assessments of a typical paper at OIST. We found that the average size was 4.8 GB, with a standard deviation of 13 GB. The data size of archived publications is summarized in Fig. 4. This number would likely be larger if we mandated archiving data that is already publicly available. Many research fields that are accustomed to larger data have repositories that can accommodate their data. For example, at OIST we have many researchers doing genomics, and these researchers

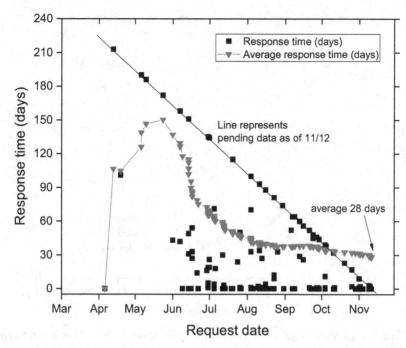

Fig. 3. Authors take about a month to respond to a request to archive data. An email is sent to authors requesting the archival of data. This graph shows the time it takes to successfully submit data or confirm that archiving is not required. The red line represents the average. The straight black line represents publications where authors have yet to resolve our request for data. (Color figure online)

often collect large sets of data. Essentially, none of this data is saved in the OIST archive because of public repositories (e.g. NCBI). There are many scientific repositories, and authors can find a relevant repository by searching a repository list (e.g. re3data.org) [7]. The data cap of 100 GB prevented very large submissions from dominating space. Only 2 out of the 79 submissions to the archive were over 80 GB, so large data submissions were not so frequent. However, large submissions may become more frequent if we set a higher data cap. We can now confirm that our original survey-based decision to cap data at 100 GB was reasonable. If needed, it may be rational to reduce the cap even further (e.g. average + 2 standard deviations ~30 GB). This is of course would be done at the expense of being more inclusive.

We would like to quantify the quality of an archive submission. Archiving data that is incomplete, poorly organized, or lacking in documentation may only have limited value. This is unfortunately difficult to objectively quantify. Consequently, the coordinator makes a subjective assessment of the submission's data and documentation. If either one was absent it is given a score of zero. Pending data is given a score of zero. If data or documentation was provided but appears to be missing requested elements it was given a score of one. If a submission was satisfactory, it was given a score of 2. The results of the results are summarized in Fig. 5. The average score is plotted as a function of time.

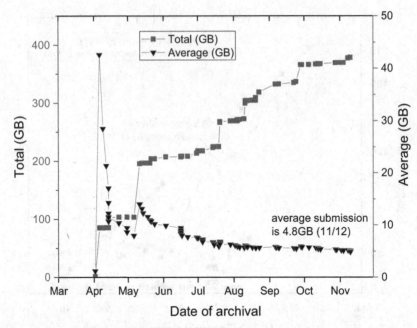

Fig. 4. The average size of a submission to the archive was about 5 GB. The left axis shows total archive data, and the right axis shows the average. (Color figure online)

The first conclusion we can make is that on average authors are willing to submit data if requested (average score above 1, most common score "satisfactory-2"). The other conclusion we can make is that authors typically are not willing to document their work (average score below 1, most common score "absent-0"). We may have been able to predict this result based on our survey, which showed that faculty were concerned about additional work burden. Typically, when writing a paper authors will organize their data for that paper. If well organized, data may not take much work to submit to the archive. Data documentation is likely something that authors do not prepare as a part of their normal routine and is likely seen as new and unwanted work.

The archive needs a high level of cooperation from authors to function. If authors are not willing to organize and submit data for archival there is no archive. Furthermore, if authors are not willing to document their work, the archive may be of limited usefulness. One of the most fundamental challenges for any institution trying to implement an archiving program is how to get researchers to participate in archival. Authors are already likely busy with their existing workload, and many expressed concerns about having to take care of the additional task of archiving data. For this reason, it is best to accommodate authors when possible. For example, at the early stages, the coordinator tried to get data into standardized formats. This was additional work that tried the limited patience of the authors and was not necessary to meet the requirement of data completeness. Accepting data as is, does potentially mean that it will be more challenging for a 3rd party to use this data. However, it is better when the data is difficult and present than simply unavailable. As can be seen from Fig. 5, simply ignoring the request for data is relatively common.

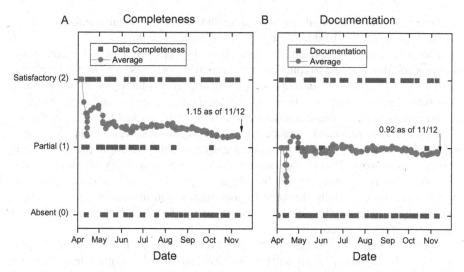

Fig. 5. On average authors are willing to submit data but are not willing to provide documentation. Every submission to the archive is given a subjective score by the coordinator. (0-absent, 1-lacking requested element, 2-Satisfactory). The red line represents the average score. (Color figure online)

Therefore, it is important to offer some form of motivation to researchers to archive their data.

Currently, OIST does a few things to motivate authors to participate in the archive. The dean of faculty affairs will investigate the unit's archive practices as part of the unit's annual review. If a faculty is seeking tenure, they may be motivated to have a good review and try to comply with archive requirements. Often the data may be organized by students and postdocs. To provide a tangible incentive to these authors, we provide a free cookie and coffee. This is a small gesture, but may provide some positive association to archiving data, which can otherwise be a rather dull task. In principle, archiving data can free up a unit's data storage space. However, the current cap on data is likely not large enough to solve the storage needs of units that generate large sets of data.

At OIST the coordinator will review the files to check for completeness, and that files reflect the data present in the paper. The simplest thing to check for is that final figure data is present and matches the publication. The next important element to check for is the presence of original data. The coordinator needs to make a judgment about what is appropriate original data for an experiment or calculation. Often original data can be a text or an excel file that is not meaningfully different from the final data. Original data may be in the form of proprietary file types (e.g., NMR measurement), and when possible, the coordinator will get the appropriate software to check these files. Sometimes the original data needs significant processing to reach its final state presented in the publication. This processing may be done by provided code or a spreadsheet. The coordinator may need to be able to look at the code and decide what was provided is likely able to produce the final data. For example, the code may analyze and plot a dataset, but the dataset was not provided along with the archive submission, and the coordinator would need to be able to spot this problem. Sometimes the coordinator can reproduce the analysis, but often it is too difficult to do in a reasonable amount of time, so these checks need to be done based on logical completeness.

Faculty commonly expressed concern about data privacy. This is reasonable, as allowing additional access to confidential data creates more opportunities for problems. In general, authors are not motivated to give broader access than they are forced to. It is possible for the coordinator to incorrectly handle access rights, and accidentally or intentionally release confidential information. Some authors asked for a non-disclosure agreement (NDA) when archiving their data. To mitigate these fears, we discussed elements of the coordinator's contract, which legally limits the use of information learned in the course of work. The standard contract is effectively redundant to a non-disclosure agreement, but faculty may need to become aware of this to put their fears aside. The data is stored in essentially the same location as research unit storage but does pass through the coordinator. It may take time to convince faculty that their data is treated with a proper level of care.

Data involving human subjects will usually require a higher level of confidentiality. Typically, participants in a study will sign a waiver that limits the scope of how their data will be used. Some authors were concerned that the archival of data would go beyond the scope that the participants agreed to. Regardless of whether this was a legitimate concern, the OIST archive chose to only keep fully anonymous human subject data. In principle, this data could be shared directly with anyone. It is kept in a private archive, but if there were to be some significant failure managing privacy it would still not violate the privacy of the subjects who participated in the study.

3   Conclusions

Research data archiving helps ensure data is available for reuse and helps with the incidence of academic fraud. However, guaranteeing high-quality data is difficult. The process requires a high level of cooperation from authors, who commonly expressed resistance to the additional work of sorting and documenting data. Authors need some motivation and trust to participate. If an archiving program does not meaningfully address these issues, the submissions may be few and of low quality. Only after receiving a proper submission is it meaningful to discuss how it can be reused. These are the fundamental issue that we experienced at the early stages of our data archiving initiative and encourage other universities to consider these points when developing a program.

References

1. Bloemers, M., Montesanti, A.: The FAIR funding model: providing a framework for research funders to drive the transition toward FAIR data management and stewardship practices. Data Intell. 2(1–2), 171–180 (2020). https://doi.org/10.1162/dint_a_00039
2. Vines, T.H., et al.: The availability of research data declines rapidly with article age. Curr. Biol. 24(1), 94–97 (2014). https://doi.org/10.1016/j.cub.2013.11.014
3. Ray, J.M. (ed.): Research Data Management: Practical Strategies for Information Professionals. DGO-Digital Original. Purdue University Press, West Lafayette (2014). https://doi.org/10.2307/j.ctt6wq34t

4. Meystre, S.M., Lovis, C., Bürkle, T., Tognola, G., Budrionis, A., Lehmann, C.U.: Clinical data reuse or secondary use: current status and potential future progress. Yearb. Med. Inform. **26**(1), 38–52 (2017). https://doi.org/10.15265/IY-2017-007
5. Meskus, M., Marelli, L., D'Agostino, G.: Research misconduct in the age of open science: the case of STAP stem cells. Sci. Culture **27**(1), 1–23 (2018). https://doi.org/10.1080/09505431.2017.1316975
6. Guidelines for Responding to Misconduct in Research. https://www.mext.go.jp/a_menu/jinzai/fusei/__icsFiles/afieldfile/2015/07/13/1359618_01.pdf
7. Homepage of re3data. https://www.re3data.org/

Symbolic Regression for Interpretable Scientific Discovery

Nour Makke, Mohammad Amin Sadeghi, and Sanjay Chawla[✉]

Qatar Computing Research Institute, HBKU, Doha, Qatar
{nmakke,msadeghi,schawla}@hbku.edu.qa

Abstract. Symbolic Regression (SR) is emerging as a promising machine learning tool to directly learn succinct, mathematical and interpretable expressions directly from data. The combination of SR with deep learning (e.g. Graph Neural Network and Autoencoders) provides a powerful toolkit for scientists to push the frontiers of scientific discovery in a data-driven manner. We briefly overview SR, autoencoders and GNN and highlight examples where they have been used to rediscover known physical phenomenon directly from data.

Keywords: Symbolic regression · Graph neural networks · Model discovery · Autoencoders

1 Introduction

Symbolic Regression (SR) is emerging as a promising machine learning tool to directly learn succinct mathematical expressions directly from data. As opposed to traditional non-linear regression methods (including deep neural networks), where the *learnt* function is often opaque and non-interpretable, SR makes it possible to deductively reason with the data-derived model. SR has already been successfully used to re-derive many known "physics" equations directly from simulated data including Newtonian mechanics [1], harmonic oscillator and Lorenz's Law [2], and opens up the possibility of inferring new physical laws that so far have resisted mathematical formalism. Recently a benchmark repository [3] for symbolic regression has been created, covering 14 contemporary SR methods and around 250 data-sets.

A relatively recent machine learning technique which complements SR is Graph Neural Networks (GNN) because of its ability to directly model space-time phenomenon and the fact it is a universal function approximator as it generalizes multilayer perceptron (MLP) neural networks. GNNs have been used in diverse applications ranging from fluid dynamics, drug discovery, climate change mitigation and adaptation and social network analysis.

Figure 1a shows an example of how symbolic regression can be used to learn a complex mathematical relationship $\frac{(x_1+x_2)}{\log(y)}$ from a data-set given in the form (x_1, x_2, y). In Fig. 1b, interactions at the atomic level can be modeled using

© Springer Nature Switzerland AG 2022
S. Sachdeva et al. (Eds.): BDA 2021, LNCS 13167, pp. 26–40, 2022.
https://doi.org/10.1007/978-3-030-96600-3_3

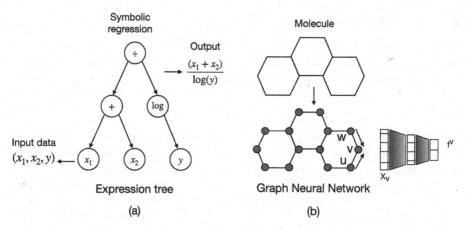

Fig. 1. (a) Symbolic Regression infers mathematical expression trees from data; (b) Atom level or Molecule level properties can be modeled using Graph Neural Networks (GNN)

GNNs. A more recent work has shown how GNNs can be combined with SR to infer basic laws of physics directly from a simulation of interacting particles [1].

The paper is organized as follows. Symbolic regression is presented in Sect. 2, where the basic approach is discussed in Sect. 2.1, test applications are presented in Sect. 2.2, and current limitations and extensions of SR are highlighted in Sect. 2.3. In Sect. 3, we review the dimensionality reduction technique using principal component analysis and autoencoders, and discuss the combination of SR with autoencoders. Finally, we review GNNs and their combination with SR in Sect. 4.

2   Symbolic Regression

Linear regression is the most conventional form of inducing models from data and is widely used due to its simplicity and effectiveness. For a given data-set $\{(y, x)_i\}_{i=1:n}$ with $y = f(x_1, x_2, x_3, ...)$, the goal is to optimize the coefficient vector Θ of a linear function $f : \mathbb{R}^n \to \mathbb{R}$ to best fit a data-set using a predefined model $f = \theta_0 + \theta_1 x$. SR generalizes the regression approach by formulating the problem as $y = f(x, \theta)$ where both the form of f and θ are unknown. Note that SR is different from neural networks because in the latter the functional form of f is also predefined, albeit more complicated than linear regression. SR induces data-driven discovery of both model structure and parameters by searching over the space of mathematical expressions the most simple and accurate expression to best fit a data-set. While the search space of mathematical expressions is discrete, the search space of parameters is continuous.

2.1 Basic Approach

The basic approach of SR used to determine the terms of the target mathematical expression is presented. Consider a data-set $(\mathbf{X}, \mathbf{Y})$ where the relationship between y_i and $\mathbf{x}$ is described by a non-linear function:

$$y = f(\mathbf{x}) \tag{1}$$

To find the mathematical expression that best fit the data vector $\mathbf{Y} \in \mathbb{R}^n$, a library $\mathcal{F}$ of k candidate functions is constructed using basic operators such as polynomials in x, trigonometric functions, exponential and logarithm, square-root, etc. The library can be further extended to include more candidate functions such that $n \gg k$, where n is the number of data inputs. A coefficient θ_i is assigned to each of the candidate functions $f_i \in \mathcal{F}$ as an activeness criterion, such that:

$$\mathbf{Y} = \mathbf{U}(\mathbf{X}) \cdot \mathbf{\Theta} \tag{2}$$

where $\mathbf{\Theta} = [\theta_0 \, \theta_1 \, \cdots \, \theta_k] \in \mathbb{R}^k$ is a vector of coefficients and $\mathbf{U} = [\mathbf{U}_1(\mathbf{X}) \, \mathbf{U}_2(\mathbf{X}) \, \cdots \, \mathbf{U}_k(\mathbf{X})] \in \mathbb{R}^{n \times k}$ is a matrix:

$$\mathbf{U}(\mathbf{X}) = \begin{pmatrix} | & | & | & | & | & | & | \\ 1 & \mathbf{X} & \mathbf{X}^2 & \mathbf{X}^3 & \sin(\mathbf{X}) & \cos(\mathbf{X}) & \exp(\mathbf{X}) & \cdots \\ | & | & | & | & | & | & | \end{pmatrix} \tag{3}$$

Each row $\mathbf{U}_i$ of this matrix is a vector of k numbers obtained by evaluating the candidate functions $\{f_j\}_{j=1}^k$ at x_i. The vector of coefficients is obtained as follows[1]:

$$\mathbf{\Theta} = (\mathbf{U}^{\mathrm{T}}\mathbf{U})^{-1} \cdot \mathbf{U}^{\mathrm{T}} \cdot \mathbf{Y} \tag{4}$$

The magnitude of a coefficient θ_i represents the weight of the contribution of the corresponding function $f_i \in \mathcal{F}$ to the result. The reconstructed data $\hat{\mathbf{Y}}$ can be evaluated using Eq. 2. Note the basic approach assumes, by construction, that the target function f is a linear combination of known functions of $\mathbf{x}$.

2.2 Applications

To illustrate a basic application of SR, we explore a simple differential equation of the form:

$$\frac{dy}{dx} = \lambda y \tag{5}$$

which yields the exact solution $y(x) = e^{\lambda x}$, for $y(0) = 1$. For $\lambda = 0.1$, a data-set $\mathbf{Y}$ is generated by evaluating the function $y(x) = e^{\lambda x}$ for a set of n values of x in the range $[0, 10]$. On this test application, a library of 7 elementary mathematical

---

[1] Technically the pseudo-inverse, U^+.

functions is constructed, e.g. $\mathcal{F} = \{1, x, x^2, x^3, \sin(x), \cos(x), \exp(x)\}$ and the coefficients vector is obtained using Eq. 4. The reconstructed function $\hat{y}(x)$ is compared to the true function $y(x) = e^{\lambda x}$ in Fig. 2 (blue curve), and a very good agreement is obtained over the full x-range with a reconstruction error of the order of 10^{-4}.

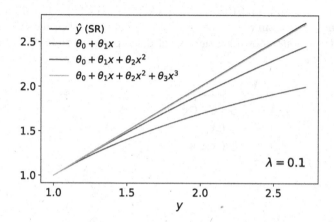

Fig. 2. Blue: reconstructed function $\hat{y}(x)$ obtained from symbolic regression versus true function $y(x) = e^{\lambda x}$. The functions $f_i \in \mathcal{F}$ with the largest contributions are shown in red, green and violet. (Color figure online)

Figure 2 also shows the selected candidate functions with the largest contributions, which are found to be $\{1, x, x^2, x^3\}$ with respective coefficients $\Theta = [1 \ 0.1 \ 0.0045 \ 0.00025]$. The SR method excludes the exponential function by a very small coefficient ($\theta_6 \sim 10^{-7}$). It identifies instead the correct terms in the Taylor expansion of $y(x) = e^{\lambda x}$ given by:

$$y(x) = e^{\lambda x} = 1 + \lambda x + \frac{\lambda^2}{2!}x^2 + \frac{\lambda^3}{3!}x^3 + \mathcal{O}(x^4) \qquad (6)$$

with coefficients $\{\theta_0 = 1, \ \theta_1 = \lambda, \ \theta_2 \approx \lambda^2/2!\}$ and $\theta_3 \neq \lambda^3/3!$ because the library includes polynomials in x up to fourth order only. Extending the library to higher order in x yields exact coefficients as expected. Note that in the simple case where the argument of the exponential is $h(x) = x$ ($\lambda = 1$), SR correctly selects e^x from the library and excludes all other candidates with null coefficients. When the argument of the exponential includes a multiplicative constant ($\lambda \neq 0$), SR excludes e^x from the library. Thus the basic approach in SR produces an exact match of the target mathematical function only when the latter belongs to the built library thus highlighting the need of an advanced version of this approach. This is particularly important in cases where the target mathematical expression is different than a linear combination of non-linear functions.

In a comparative perspective, the LASSO model was applied on the same input data $\mathbf{Y}$. LASSO is a linear model that estimates sparse coefficients by including an ℓ_1-regularization term on the coefficients Θ into the optimization problem:

$$\Theta = \arg\min_{\Theta} \|\hat{\mathbf{Y}} - \mathbf{Y}\|_2 + \alpha\|\Theta\|_1 \tag{7}$$

where α weights the sparsity constraint. For $\alpha = 0.1$, the resulting coefficient vector is $\Theta = [0.0 \quad 0.0 \quad 0.01987 \quad -0.00027 \quad -8.5310^{-6} \quad 0.0 \quad 0.0]$ selecting the functions $\{x^2, x^3, \sin x\}$ as active. Figure 3 compares the reconstructed and the true distributions, showing a significant discrepancy both in numerical values and x-dependence.

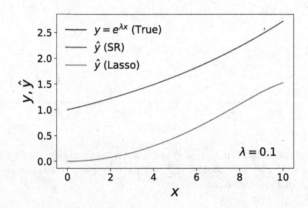

Fig. 3. Reconstructed function $\hat{y}(x)$ obtained from symbolic regression (red) and LASSO (green) versus x, compared to true distribution $y(x) = e^{\lambda x}$ (blue). (Color figure online)

We explore in a different application an observable in physics that represents the probability of a given process to occur. For example, consider in high-energy physics the scattering of two particles, an electron (e^-) and a positron (e^+), that annihilate to create another pair of particles. In the center-of-mass frame, the two particles collide head-on and the two outgoing particles are produced in opposite direction, where θ is the angle of the outgoing particle μ^- with respect to the incoming electron direction, as shown in Fig. 4. The observable that represents this collision can be derived [4]:

$$\frac{dy}{d(\cos\theta)} = C(1 + \cos^2\theta) \tag{8}$$

where C is a energy-dependent term. At fixed energy, a dataset $\mathbf{Y}$ can be generated using $Cdy/d(\cos\theta) = 1 + \cos^2\theta$ for n values of $\theta \in [0, \pi]$. The SR method described above is applied with different bases libraries, including polynomial basis functions, trigonometric functions, exponential and logarithm functions, and finally a combination of all: $\mathcal{F}_1 = \{1, \theta, \theta^2, \theta^3, \theta^4, \theta^5\}$, $\mathcal{F}_2 = \{\cos\theta, \sin\theta, \cos^2\theta, \sin^2\theta\}$, $\mathcal{F}_3 = \{e^\theta, e^{-\theta}, \log\theta\}$ and $\mathcal{F}_4 = \{\mathcal{F}_1, \mathcal{F}_2, \mathcal{F}_3\}$.

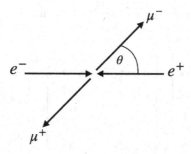

Fig. 4. The scattering process $e^- e^+ \to \mu^- \mu^+$ viewed in the center-of-mass frame. The electron and the positron annihilate to produce a new pair of particles [4].

The true $(y(\theta))$ and the reconstructed $(\hat{y}_i(\theta))$ data obtained for different libraries $\mathcal{F}_i$ are compared in Fig. 5a as a function of $\cos\theta$, and the ratios $\hat{y}_i/y$ are shown in Fig. 5b. An excellent agreement is obtained for $\mathcal{F}_1$ and $\mathcal{F}_2$ and a significant discrepancy is observed for $\mathcal{F}_3$ and $\mathcal{F}_4$.

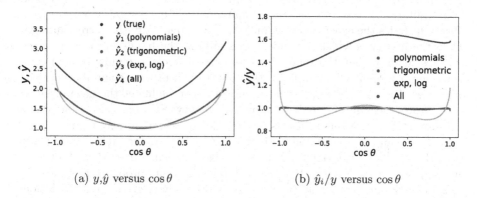

(a) $y, \hat{y}$ versus $\cos\theta$ (b) $\hat{y}_i/y$ versus $\cos\theta$

Fig. 5. (a) True $(y(\theta))$ and reconstructed $(\hat{y}_i(\theta))$ data versus $\cos\theta$ and (b) the ratios $\hat{y}_i/y$ ($\mathcal{F}_1 \equiv$ polynomials, $\mathcal{F}_2 \equiv$ trigonometric, $\mathcal{F}_3 \equiv$ exp, log and $\mathcal{F}_4 \equiv$ all). The green curve supersede both blue and red curves in (a). (Color figure online)

For both cases $\mathcal{F}_1$ and $\mathcal{F}_2$, the coefficients are shown in Table 1. In the case of a pure polynomial library $\mathcal{F}_1$, the construction is excellent (within $<1\%$), and higher orders of polynomials in θ leads to a perfect agreement ($<1\%_0$) between $\hat{y}$ and y. However the resulting coefficients Θ do not match with the correct terms in the Taylor expansion of $y(\theta) = 1 + \cos^2\theta$. In the case of a purely trigonometric library $\mathcal{F}_2$, the correct terms are identified yielding $\hat{y}(\theta) = 2\cos^2\theta + \sin^2\theta = 1 + \cos^2\theta$, i.e. $\hat{y}(\theta) = y(\theta)$.

Table 1. Coefficients vectors $\Theta_{1,2,3}$ obtained for the libraries $\mathcal{F}_{1,2,3}$ for the test application of Eq. 8.

$$
\begin{bmatrix}
\mathcal{F}_1: & \Theta_1 \\
1: & 2.0000 \\
\theta: & -0.0395 \\
\theta^2: & -2.1826 \\
\theta^3: & 1.3084 \\
\theta^4: & -0.2076 \\
\theta^5: & -0.0001
\end{bmatrix}
\begin{bmatrix}
\mathcal{F}_2: & \Theta_2 \\
\cos\theta: & 0.00 \\
\sin\theta: & 0.00 \\
\cos^2\theta: & 2.00 \\
\sin^2\theta: & 1.00
\end{bmatrix}
\begin{bmatrix}
\mathcal{F}_3: & \Theta_3 \\
\exp(\theta): & 0.0994 \\
\exp(-\theta): & 2.5274 \\
\log\theta: & 0.0551
\end{bmatrix}
$$

For the extended library $\mathcal{F}_4$, the SR method does not reproduce either the numerical values of the data distribution or its θ-dependence.

2.3   Limitations and Extensions

The applications discussed in Sect. 2.2 show that the basic approach of SR suffers from limitations to extract the target mathematical expression unless the built library includes the exact function, and struggles to identify the correct function when the library is widely extended. In addition, the basic approach assumes, by construction, that the target function f is a linear combination of known functions of **x**, limiting its application to general SR problems and its success to only special cases. The arguments discussed above highlight the need of either an improved version of this basic approach or new SR approaches, for use on real-world data-sets where the level of complexity is significantly high. Existing approaches can be classified into two categories: (a) combine SR and deep learning (AEs, GNNs); (b) group a collection of algorithms into one SR method. Such new approaches are preferably applied in physics where problems are governed by symmetries and conservation laws, and data-sets have ground truth solutions allowing to check whether the learned models match with analytical solutions.

The method proposed in [5], entitled *AI Feynman*, is an advanced version of symbolic regression. It takes into consideration that physical problems could be simplified due to properties underlying functions have in physics such as symmetry, factorization, and dimensionality constraints. *AI Feynman* consists of a chain of modules (e.g. polynomial fit, brute-force search, etc.) that exploit physics-inspired constraints, combined with a neural network. The modules are applied in turn and in case of failure, the neural network is trained to discover fundamental properties of the underlying function to simplify the problem by reducing the number of independent variables. For example, symmetry is tested by checking if f depends upon two input variables through their difference and a new single variable is created when it's the case. This task reduces the number of independent input variables, and is repeated for all basic operations $(+, -, *, /)$ and for all pairs of input variables. f is also tested for factorization, i.e. $f(x_1, x_2, x_3) = g(x_1)h(x_2, x_3)$, and the dimensionality constraint which requires the units of the two sides of an equation to match is applied. The method uses a large database which was created using 100 equations covering different physics areas from Feynman's book, with an extra 20 famous equations selected for being complicated, and excluding equations involving derivatives or integrals. Three simple examples are shown in Eq. 9. The *AI Feynman* shows better performance over the state-of-the-art methods with 100% of exact solutions out of the 100 equations versus 71% by the commercial software *Eureqa*. Further details can be found in [5].

$$\text{Distance:} \quad d = \sqrt{(x_2 - x_1)^2 + (y_2 - y_2)^2}$$
$$\text{Gravitational force:} \quad F = \frac{Gm_1 m_2}{(x_2 - x_1)^2 + (y_2 - y_1)^2 + (z_2 - z_1)^2} \quad (9)$$

AI Feynamn was applied to the data simulated in the applications described above. For the first application (Eq. 5), the exact solution $y(x) = e^{(0.1)x}$ was found in addition to its Taylor expansion approximation $y(x) = 1 + (0.1)x$, and for the second application (Eq. 8) the algorithm extracted the solution $\sin^{-1}\left(0.25.10^{-10} + \sin(\cos^2\theta + 1)\right)$ which equals $(1 + \cos^2\theta)$.

Although this model shows significant improvements compared to state-of-the-art methods on ground-truth physics problems, thanks to the power of NN, it poorly performs on real-world data-sets where the noise level is expected to be significantly higher than in the almost noise-free simulated data as discussed in the benchmarking platform [3]. In addition, the enormous surge of collected data in different areas where symbolic regression is meant to have a strong impact calls for use of specific machine learning tools, such as dimensionality reduction tools. More recent methods learn symbolic expression trees (like in Fig. 4(a)), by combining RNNs with Reinforcement Learning approaches [6].

3 Dimensionality Reduction

High-dimensional data-sets play a key role in pushing the boundaries in machine learning models as they offer more features about the internal data generating process. Dimensionality reduction plays a key role in removing redundant and irrelevant features while keeping the "useful" part of the information in the reduced representation. Dimensionality reduction is the transformation of data from a high-dimensional space into a lower-dimensional space and it is likely to play an important role in conjunction with SR to create interpretable models.

3.1 Principal Component Analysis (PCA) and Autoencoders

Principal Component Analysis (PCA) is one of the most popular dimensionality reduction techniques. The basic idea of PCA is that while many data sets are high-dimensional, they tend to inhabit a low-dimensional manifold. PCA operates by projecting data into a lower-dimensional representation through a linear transformation that searches the directions of maximal variance of the data, as shown in Fig. 6 where the vectors shown are the eigenvectors of the data's covariance matrix, scaled by the square root of the eigenvalues. Instead of applying on the original data set, SR can be applied to the data set transformed with PCA. By using the transformed dimensions, SR can generate succinct representations and furthermore by focusing on dimensions with higher eigenvalues, it can systematically deal with noise elimination. In the next section we will review deep autoencoders which are a nonlinear generalization of PCA and have been used in conjunction with SR to rediscover the dynamics of many physical systems.

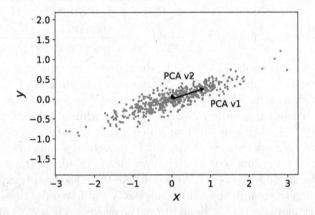

Fig. 6. PCA applied to data sample. The vectors are the eigenvectors obtained using PCA fit to the data, scaled by the square root of their corresponding eigenvalues. PCA can be combined with SR to create succinct models in the transformed and reduced space. (Color figure online)

Mathematically, PCA can be represented as follows. Suppose the data matrix $\mathbf{X} \in \mathbb{R}^{N \times D}$ has zero mean, Then PCA can be interpreted as the result of the factorisation [7]:

$$\min_{\mathbf{W}^T\mathbf{W}=\mathbf{I},\mathbf{Z}} \|\mathbf{X} - \mathbf{W}\mathbf{Z}\|_F^2 = \min_{\mathbf{U}} \|\mathbf{X} - \mathbf{X}\mathbf{U}\mathbf{U}^T\|_F^2. \tag{10}$$

Here, the reconstruction matrix is $\hat{\mathbf{X}} = \mathbf{X}\mathbf{U}\mathbf{U}^T$, where $\mathbf{U} \in \mathbb{R}^{D \times K}$ for some number of *latent dimensions* $K \ll D$. $\mathbf{X}\mathbf{U}$ the encoding (projection) of $\mathbf{X}$ into a K-dimensional subspace, with $\mathbf{U}^T$ as the decoding (inverse projection) back into the original D dimensional space.

The underlying assumption of PCA, that a linear subspace can explain the data can be relaxed by using a non-linear activation function. For example, consider the factorization:

$$\min_{\mathbf{U},\mathbf{V}} \|\mathbf{X} - f(\mathbf{X}\mathbf{U})\mathbf{V}\|_F^2 \tag{11}$$

for a non-decreasing *activation function* $f : \mathbb{R} \to \mathbb{R}$, and $\mathbf{U} \in \mathbb{R}^{D \times K}, \mathbf{V} \in \mathbb{R}^{K \times D}$. Equation 11 corresponds to an autoencoder with a single hidden layer [8]. Some popular examples of $f(\cdot)$ include the sigmoid $f(a) = (1 + \exp(-a))^{-1}$ and ReLU $f(x) = \max(0, a)$. As in the case of PCA, we can interpret $\mathbf{XU}$ as an encoding of $\mathbf{X}$ into a K-dimensional subspace; however, by applying a nonlinear $f(\cdot)$, the data is projected onto a nonlinear manifold.

3.2   SR Combined with Autoencoders

This section discusses a recent method [2] that combines SR, sparse methods and autoencoders to recover the dynamics of some well known physical systems from time series data. The dynamics of systems with state $x(t)$ at time t are governed the differential equation:

$$\frac{d}{dt}x(t) = f(x(t)) \tag{12}$$

The proposed approach in [2] relies on searching for a reduced coordinate system, using an autoencoder, where the dynamics have a simpler representation. The equation that describes the dynamics is then associated with the reduced coordinates as follows:

$$\frac{d}{dt}z(t) = f(z(t)) \tag{13}$$

Here $z(t)$ represents the autoencoder's latent variable (the output of the encoder). The novelty of the approach is that the lower-dimensional representation of the autoencoder, the (sparsity regularized) symbolic regression parameters are learnt simultaneously by creating a customized loss function which is a sum of the autoencoder loss, the SR loss and the ℓ_1 regularizer on the SR parameters.

A basic example to mention is a nonlinear pendulum whose dynamics are governed by a second-order differential equation given by $\ddot{x} = -\sin x$. The data consists of a simulated video of a pendulum, which is used to generate a series of snapshot images in a two-dimensional spatial representation. After training, the method extracted the equation $\ddot{z} = -0.99\sin z$ for the latent variable, and thus correctly discovered the dynamics of the system through a reduced representation.

The use of autoencoder as a powerful dimensionality reduction tool has significantly pushed the boundaries of model discovery for high-dimensional data, surpassing a major limitation of prior methods while prioritizing the interpretability of the learned model. However the requirement of noise-free data-sets remain the main current limitation in the discussed approaches [2,5].

4   Graph Neural Networks

Graph neural networks (GNNs) [9] are a class of deep neural networks designed to solve real-world problems that have graph-like structured data. A graph describes the relationship between a collection of entities as illustrated in

Fig. 7(a). It is defined by a set of nodes (also called vertices) representing the objects, $V = \{v_i\}_{i=1:N_v}$ where N_v is the number of nodes and each v_i is a node's vector of attributes, and a set of edges (also called links) that define the relations between the objects, $E = \{e_{ij}\} \in V \times V$ where each $\mathbf{e}_{ij}$ is an edge's vector of attributes and i and j are the indices of the connected nodes, as summarized in Fig. 7(b). The entire graph $G = (V, E)$ is defined by a global attribute $\mathbf{u}$ and is represented by an adjacency matrix.

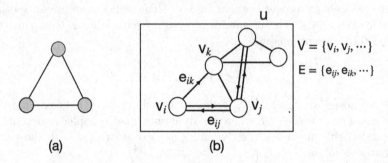

(a)   (b)

Fig. 7. (a) A graph of three nodes and three edges. (b) Nodes and egdes are represented by vectors of attributes v and e respectively, and the entire graph is represented by a global attribute u. Edges are directed if dependencies between nodes are directional.

The concept of a graph is applicable to nearly all types of real-world problems such as physical and chemical systems, images, text sequences, social networks, traffic networks, brain networks, etc., with some are shown in Fig. 8 as examples. Figure 9 shows their assigned graph structures, where nodes and edges are defined in each case: (a) the particles m_1 and m_2 of a physical system define the nodes and their interaction defines the edge; (b) in chemistry, the graph structure of a molecule consists of the atoms which define the nodes and the chemical bonds which define the edges; (c) each pixel of an image defines a node and its connections with adjacent pixels define its edges; (d) in a text sequence, each word defines a node and its connection to the following word defines an edge, (e) people in a social network define the nodes and their contacts define the edges, (f) sensors of a traffic network define the nodes and their relative distance define the edges.

In GNN, each node is represented by a vector of attributes of dimension D_v, $\mathbf{v}_i \in \mathbb{R}^{D_v}$, and each edge connecting two nodes of indices i and j is represented by a vector of attributes and directions of dimension D_e, i.e. $\mathbf{e}_{ij} \in \mathbb{R}^{D_e}$. GNN uses a multilayer perceptron (ϕ) with L hidden layers for each of its components (nodes, edges, entire-graph) and operates in a consecutive way from the edge, to the node, to the graph-level.

Edge-Level: The edge function ϕ^e is a multilayer perceptron. It is applied for each edge, takes the attributes of the connected nodes (i, j) as inputs and computes an updated edge attribute $\mathbf{e}'_{ij} = \phi^e(\mathbf{e}_{ij}, \mathbf{v}_i, \mathbf{v}_j) \in \mathbb{R}^{D_e}$ for every hidden

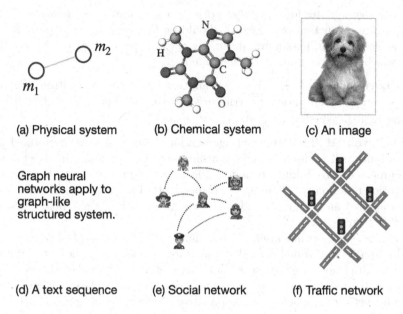

(a) Physical system   (b) Chemical system   (c) An image

Graph neural
networks apply to
graph-like
structured system.

(d) A text sequence   (e) Social network   (f) Traffic network

Fig. 8. Examples of real-world systems that have a graph-like structure.

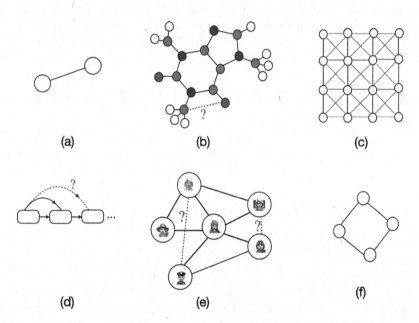

(a)   (b)   (c)

(d)   (e)   (f)

Fig. 9. Graph structures assigned to the examples in Fig. 8. The nodes are: (a) the particles in a physical system, (b) the atoms of a molecule, (c) the pixels of an image, (d) the words of a text sequence, (e) people in a social network, (f) sensors in a traffic network, and the edges represent their relationship.

layer. All edge updates for edges that project to node i are then aggregated using an aggregation function $\rho^{e \to v}$, which should be invariant to input permutation (such as summation, maximum, mean, etc.), e.g. $\bar{\mathbf{e}}'_i = \rho^{e \to v}(\mathbf{e}_{ij}) = \sum_j \mathbf{e}_{ij}$. The latter will be used as input in the next step.

Node-Level: The node function ϕ^v is another multilayer perceptron. For each node, it computes an updated attribute $\mathbf{v}'_i$, taking as input the node attribute and the aggregated edge attribute, e.g. $\mathbf{v}'_i = \phi^v(\bar{\mathbf{e}}'_i, \mathbf{v}_i) \in \mathbb{R}^{D'_v}$.

Graph-Level: At this level, two aggregations operations are performed and their outputs will be used in the following update operation. All edge updates are aggregated into $\bar{\mathbf{e}}$ and all node updates are aggregated into $\bar{\mathbf{v}}$ using the aggregation functions $\rho^{e \to u}$ and $\rho^{v \to u}$ respectively. The entire-graph function ϕ^u is then applied on the graph and computes an update of the global attribute, $\mathbf{u}' = \phi^u(\bar{\mathbf{v}}, \bar{\mathbf{e}}, \mathbf{u})$.

Finally, a joint optimization of all components of a GNN (nodes, edges and global-graph) is performed such that the graph is invariant under permutation.

GNN can perform an edge-level, node-level and graph-level prediction tasks, which can be a simple number, a vector of features, or a classification output. A graph-level task predicts a single property of the entire graph such as graph classification, examples of this application are image classification (Fig. 8(c)) and sentiment analysis (Fig. 8(d)). A node-level task predicts a node attribute or a single property for each node in a graph, and an edge-level task predicts the connection between two objects in a graph such as link prediction (Fig. 8(b, d, e)). Table 2 lists examples of possible prediction tasks for some of the systems illustrated in Fig. 8.

Table 2. Predictions tasks at the node, edge, and entire-graph levels for some of the systems illustrated in Fig. 8.

				Graph neural networks apply to graph-like structured system.
node-level	Velocity/energy of a particle	Energy of an atom	Image segmentation	prediction of the next word
edge-level	Interaction type or distance between particles	Inter-atomic potential type or distance between atoms	link prediction	link prediction
graph-level	Kinetic energy of the system	Energy of the system	Image classification	Sentiment analysis

4.1 SR Combined with Graph Neural Network

Cranmer et al. [1] have presented an approach where GNNs are used to model data derived from simulated physical systems followed by SR to extract interpretable physical relationships which coincide with the underlying dynamics. Figure 10 summarizes their methodology.

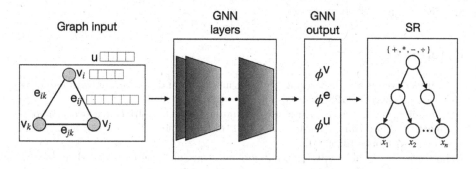

Fig. 10. A summary of the approach in [1] that combines SR and GNN. The structure of the graph is first defined, GNN is then trained on data, and the representation learned by GNN components is fed into a SR algorithm.

A case study is the Newtonian dynamics which describe the dynamics of particles in a system according to Newton's laws of motion. The data-set consists of a N-body system with known interaction (force law F such as electric, gravitation, spring, etc.), where particles (nodes) are characterized by their attributes (mass, charge, position, velocity and acceleration) and their interaction (edges) are assigned attribute of dimension 100. The GNN functions are trained to predict instantaneous acceleration for each particle using the simulated data and then applied on a different data sample. The study shows that the most significant edge attributes, say $\{e_1, e_2\}$, fit to a linear combination of the true force components, $\{F_1, F_2\}$, which were used in the simulation showing that edge attributes can be interpreted as force laws. The most significant edge attributes were then passed into a SR algorithm that could recover many analytical expressions that are equivalent to the simulated force laws. The proposed approach was also applied to data-sets in the field of cosmology, and it discovered potentially new laws for cosmological dark matter.

5   Conclusion

In this paper we have reviewed symbolic regression (SR) along with some applications which highlight both its strengths and weaknesses. In the most basic form, if the dependent variable is a linear combination of functions from a predefined library, the basic approach is able to recover the underlying relationship that generated the data. More complex algorithms that model the relationship between the data variable as an expression tree have been recently introduced and have shown initial promise to recover nonlinear functional relationships. SR techniques have successfully been combined with autoencoders to recover well known physical phenomenon including the equation of the pendulum and Lorenz's equations. A more recent work have applied SR to extract interpretable relationships from graph neural networks. The field of SR is in its infancy and as the need for interpretable AI increases, the demand for more sophisticated SR algorithms is likely to grow.

References

1. Cranmer, M., et al.: Discovering symbolic models from deep learning with inductive biases. arXiv:2006.11287 [cs.LG]
2. Champion, K., Lusch, B., Kutz, J.N., Brunton, S.L.: Data-driven discovery of coordinates and governing equations. PNAS **116**(45), 22445–22451 (2019). https://doi.org/10.1073/pnas.1906995116
3. La Cava, W., et al.: Contemporary symbolic regression methods and their relative performance. In: NeurIPS 2021 Datasets and Benchmarks Track (Round 1) (2021)
4. Halzen, F., Martin, A.D.: Quarks and Leptons: An Introductory Course in Modern Particle Physics. Willey, Hoboken (1984)
5. Udrescu, S.M., Tegmark, M.: AI Feynman: a physics-inspired method for symbolic regression. Sci. Adv. **6**(16), eaay2631 (2020). https://doi.org/10.1126/sciadv.aay2631
6. Petersen, B.K., Larma, M.L., Mundhenk, T.N., Santiago, C.P, Kim, S.K., Kim, J.T.: Deep symbolic regression: recovering mathematical expressions from data via risk-seeking policy gradients. In: International Conference on Learning Representations (2021)
7. Bishop, C.M.: Pattern Recognition and Machine Learning. Springer, New York (2006)
8. Goodfellow, I., Bengio, Y., Courville, A.: Deep Learning. MIT Press, Cambridge (2016)
9. Battaglia, P.W., et al.: Relational inductive biases, deep learning, and graph networks. ArXiv:1806.01261 (2018)

A Front-End Framework Selection Assistance System with Customizable Quantification Indicators Based on Analysis of Repository and Community Data

Koichi Kiyokawa[ID] and Qun Jin[✉][ID]

Graduate School of Human Sciences, Waseda University, Tokorozawa, Saitama 3591192, Japan
koichi19937@toki.waseda.jp, jin@waseda.jp

Abstract. Front-end frameworks are in increasing demand in web application development. However, it is difficult to compare them manually because of their rapid evolution and big variety. Prior research has revealed several indicators that developers consider important when selecting a framework. In this study, we propose and develop a system that assists developers in the selection process of a front-end framework, which collects data from repository and user community, such as GitHub and other sources, and quantifies a set of indicators. This system is built as a web application, and users can specify the importance of an indicator by adjusting the weight for each indicator. As a result of semi-structured interviews with front-end developers after using our system based on practical scenarios, we found that the proposed system is effective for narrowing down the framework and has practicality.

Keywords: Technology selection · Front-end framework · Repository mining · Semi-structured interview

1 Introduction

In web application development, to efficiently create complex UIs, more and more developers use front-end frameworks, such as React, Vue, and Angular, etc. However, it is difficult to compare them manually because they evolve rapidly and there are many types of frameworks. Some prior research has revealed that there are several indicators that developers consider important when selecting a framework. In this study, to assist developers in the selection process, we propose and develop a system that quantifies indicators by collecting and analyzing data from GitHub and other sources. The system is built and provided as a web application. It allows users to specify the importance of each indicator by setting weights to deal with their various preferences.

The remainder of this paper is organized as follow. Section 2 presents related work, such as studies that interviewed developers to reveal important factors in selecting a framework, and existing framework comparison systems. In Sect. 3, we describe the indicators, algorithms, and implementation and UI of the proposed system. Section 4

© Springer Nature Switzerland AG 2022
S. Sachdeva et al. (Eds.): BDA 2021, LNCS 13167, pp. 41–55, 2022.
https://doi.org/10.1007/978-3-030-96600-3_4

discusses the evaluation experiment, and results. Section 5 provides the conclusion and highlights the future work.

2 Related Work

2.1 Criteria for Selecting Third-Party Libraries

Larios et al. studied the factors that influence the selection process of third-party libraries [1]. Sixteen developers from 11 different industries were interviewed, and 115 developers involved in library selection were surveyed. As a result, they concluded that there are three major factors in library selection: technical, human, and economic factors. The human factor and economic factor are mostly qualitative. On the other hand, technical factors include functionality (size and complexity, whether it meets the purpose), quality (ease of use, security, performance, test coverage), release (whether it is maintained, maturity and stability, development activity), and community (popularity, information sharing activity), and there is a possibility to make quantitative judgment. Gizas et al. [2] and Graziotion et al. [3] also interviewed developers and revealed indicators, such as activity of the community, usability, and reputation.

2.2 Comparison of Specific Frameworks

Kaluza et al. compared the syntax, learning difficulty, and performance of three front-end frameworks: Angular, React, and Vue [4]. This study compared specific versions of specific frameworks. As new versions of the libraries are released, the comparison results become outdated.

2.3 Existing Library Comparison System

Npm is a registry that stores libraries for JavaScript. Npms is a comparison tool for libraries published on npm [5], which obtains data such as the number of downloads from npm and source code and commit information from GitHub, and scores libraries based on three indicators: quality, popularity, and maintenance. The total score is calculated by Eq. (1).

$$TotalScore = 0.3 \times Quality + 0.35 \times Popularity + 0.35 \times Maintenance \quad (1)$$

The coefficients such as 0.3 and 0.35 in Eq. (1) are arbitrary values determined by the developers of Npms. It is expected that there will be differences in which indicators are important to each user of the system, but Npms is not able to deal with such differences. The user interface of the system is designed to allow users to enter the library name or keywords registered in the library in the search field, and search by partial matching. The keywords can be freely set by the creator of the library, and there is no unified standard. For this reason, it is difficult to compare libraries in the same category, such as front-end frameworks.

Orion is a software search system that uses a unique query language [6]. It is designed not only for selecting libraries, but also for finding reusable code and open-source projects that can contribute to development. The system uses an original query language to search for "projects developed by less than 10 people, containing more than 10 million lines of code, and with a bug fix rate of more than 80%". While the high degree of freedom of the search makes it possible to respond to the detailed requests of users, it is costly to learn a proprietary query language, and users must think of complex queries even to calculate a single indicator.

3   An Assistance System with Customizable Indicators Based on Related Data Analysis

Unlike the comparison of specific versions conducted by Kaluza et al. [4], in this study the system is designed to allow comparison based on the latest data of front-end frameworks. In addition, unlike Npms [5], our proposed system allows users to set how much important each indicator is.

3.1   Indicators

In this study, we use the indicators shown in Table 1, referring to related works [1–3].

Table 1. Indicators and data used for analysis and quantification

Indicator	Data used for analysis and quantification
Development activity	Git commit history
Information sharing activity	Issues in GitHub, question in stack overflow
Similarity of APIs with a specific framework	Codes from an open-source project called "Realworld"
Maintenance activity	Issues in GitHub
Maturity	Version release history
Popularity	Star history from GitHub, download history from npm

In common with each indicator, the policy is to increase the weight of newer data. To achieve this, Eq. (2) is used to calculate the score so that the larger the number of days elapsed, the smaller the value.

$$score_{age}(d) = \frac{D}{D + d} \tag{2}$$

where d is the number of days elapsed for the data, and D is a constant. This means that the importance of the data from D days ago is half of the data from today. In this study, we set $D = 30$. And the data to be obtained is limited to the most recent one year. This is because increasing the number of years would give too much advantage to frameworks with older release dates. Also, each indicator is normalized to the range of zero to one, and then presented to the users.

The definitions of six indicators and their quantification methods are described as follows.

Development Activity
This indicator shows the degree to which new features are added and bugs are fixed. By selecting a front-end framework that actively provides these services, when a front-end developer notices a lack of functionality or a bug in the framework and reports it, the probability of having the functionality added or the bug fixed in a relatively short period of time is increased. This is expected to improve the efficiency of front-end development. Framework developers store changes to the source code in commit-meaningful units. Therefore, the more commits with a large number of changed lines are made, the more active the development is considered to be. In addition, the more recent the commit is, the higher the value is set. Based on the above, we define the development activity as in Eq. (3).

$$score_{dev} = \sum_{}^{commits} n_{lines} \times score_{age}(d_{commit}) \tag{3}$$

where n_{lines} is the number of changed lines in the commit, and d_{commit} is elapsed days since the commit was made.

Information Sharing Activity
This indicator shows the extent to which the questions and answers about the framework are being asked and answered. For example, posting issues on GitHub (e.g., bug reports, usage questions, etc.), asking questions on Stack Overflow, and answering those questions. The higher the number of comments on an issue, the higher the value; and the higher the number of answers on Stack Overflow, the higher the value. Based on these, we define this indicator as in Eq. (4))

$$score_{info} = \sum_{}^{issue\ comments} c_{character} \times score_{age}(d_{comment})$$
$$+ \sum_{}^{questions} c_{answer} \times score_{age}(d_{question}) \tag{4}$$

Where $c_{character}$ is the character count in each issue comment $d_{comment}$ is elapsed days since commented c_{answer} is the answer count in each question. $d_{question}$ is elapsed days since the question was created.

Similarity of APIs with a Specific Framework
This indicator shows the degree of code similarity between the implementations of the same feature in the two frameworks. Figure 1 shows two examples of the codes for implementing the same counter in Vue and Svelte as front-end frameworks. The similarity of these codes is an example of this indicator. One of the use cases for this indicator is searching for a framework that is similar to the one already learned to reduce the learning cost.

```
1    <template>
2      <div>
3        <span>{count}</span>
4        <button>+1</button>
5      </div>
6    </template>
7
8    <script>
9    export default {
10     data: () => ({
11       count: 0
12     }),
13     methods: {
14       increment() {
15         this.count++
16       }
17     }
18   }
19   </script>
```

```
1    <script>
2      let count = 0
3      function increment() {
4        count++
5      }
6    </script>
7
8    <div>
9      <span>{count}</span>
10     <button>+1</button>
11   </div>
```

(a) Vue   **(b)** Svelte

Fig. 1. Code of implementing a counter with Vue and Svelte

In this study, we calculated the similarity of APIs using the source code of Real-World open source [7], which is an open-source project that provides implementations of blog applications with certain specifications by various frameworks. The purpose of the project is to provide a practical demo application. It is also suitable for comparing codes between frameworks. We used the source code of the login page for comparison. This is because the granularity of component, request method with the backend, and state management method differ among frameworks for the article list page, etc. It means that the differences other than API are too large. As for the login page, it is simple enough that the main difference is the API, and it is suitable for comparison. The following algorithm was used to calculate the similarity between two source codes A and B.

1. As preprocessing, remove stop words (such as semicolon, indent, and line breaks) from A and B.
2. Split each of A and B with a bigram (two adjacent strings).
3. Combine the elements created in 2 into a set, make a list by sorting the elements in this set in lexical order.
4. Vectorize each of A and B by counting the number of times an element appears in the order of the list of 3.
5. Calculate the cosine similarity of the two vectors created in Step 4.

The reason for the sorting in Step 3 is to make sure that the order in which the codes appear does not reduce the similarity. For example, the source codes in Figs. 1 (a) and (b) have similar APIs, but the order of the template and script tags are reversed.

Maintenance Activity

This indicator shows the extent to which bug reports and specification questions are resolved. A well-maintained system is one in which issues are resolved quickly, and in which development members comment on the issues frequently. Conversely, the longer an issue is left unresolved, the less maintained it is. In particular, an issue with a large number of comments means that many users are suffering from the same problem, and the negative impact of neglect is significant. Based on these, we divide them into the following three sub scores.

1. *Speed of resolving issues*

The shorter the number of days between when the issue is opened and when it is closed, the larger the value, and the newer the issue, the larger the value. We define this sub score as in Eq. (5).

$$subscore_{resolve\_speed} = \sum^{resolved\ issues} score_{age}(d_{elapsed}) \times score_{age}(d_{resolve}) \quad (5)$$

where $d_{elapsed}$ is elapsed days from when the issue was created to resolve it. $d_{resolve}$ is elapsed days from when the issue was resolved to today.

2. *Degree of development members' participation to the issue.*

For comments posted by members on an issue, the more characters in the comment and the more recent the comment, the larger the value. We define this sub score as in Eq. (6).

$$subscore_{comment} = \sum^{issue\ comments}_{by\ maintenars} c_{character} \times score_{age}(d_{comment}) \quad (6)$$

where $c_{character}$ is the character count in each issue comment, and $d_{comment}$ is the elapsed days since the comment was created.

3. *Degree of neglecting the issue.*

The longer an issue is left with many comments, the larger the value. In contrast to 1 and 2, the smaller is better. We define this sub score as in Eq. (7).

$$subscore_{neglect} = \sum^{opened\ but}_{unresolved\ issues} c_{character} \times \frac{1}{score_{age}(d_{open})} \quad (7)$$

where $c_{character}$ is the character count in each issue comment, and d_{open} is elapsed days since the issue was opened. Note that we use the reciprocal because the value of $score_{age}$

decreases as the elapsed days increases, but in this case, we make this sub score to increase as the elapsed days increases.

Finally, we define maintenance activity score as in Eq. (8).

$$score_{maint} = N\left(subscore_{resolvespeed}\right) + N\left(subscore_{comment}\right) - N\left(subscore_{neglect}\right) \quad (8)$$

where N is a function that normalizes the argument to the range zero to one.

Maturity

This indicator shows the degree of stability with few destructive changes. We define this indicator from the publication history of a version, referring to Gizas et al. [2]. Figure 2 shows the maturity level classified by the release history of the versions. The circles in the figure indicate the events when each version was released. In Fig. 2(a), the period when new versions were actively released has passed, the specifications have been finalized, and the interval between the release of versions has gradually become longer. Therefore, it can be said that this is a stable state. Figure 2(b) shows a state in which the latest version is actively released and is not stable. In Fig. 2(c) the version was actively released in the past, but the development was stopped before it became mature. When the development resumes, destructive changes may occur.

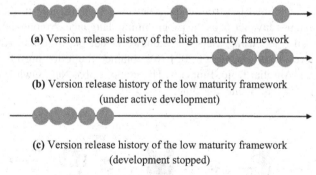

(a) Version release history of the high maturity framework

(b) Version release history of the low maturity framework
(under active development)

(c) Version release history of the low maturity framework
(development stopped)

Fig. 2. Version release history and maturity

Based on the above, the maturity will become larger when the number of days between each version and the latest version is larger. Therefore, it is defined as in Eq. (9).

$$score_{maturity} = \sum^{released\ versions} \frac{1}{score_{age}\left(d_{diff}\right)} \quad (9)$$

where d_{diff} is the difference in days between the latest version and each version.

Popularity

This indicator shows how much attention is paid by front-end developers. With reference to Npms [5], in this study, the more recent the download history of npm and the star history on GitHub, the higher the value. It is important to note that the number of downloads tends to increase exponentially as the popularity of the framework increases. As the popularity increases, more and more projects adopt the framework. In fact, developers are not the only ones who download libraries. For example, downloads also occur when automated tests are run by CI (continuous integration) [8]. In other words, as the popularity of frameworks increases, these non-developer downloads will increase. Therefore, downloads need to be evaluated on a logarithmic scale. Based on the above, the popularity is defined as in Eq. (10)

$$score_{popularity} = \sum_{d=0}^{365}(C_{star}(d) + \log C_{download}(d)) \times score_{age}(d) \qquad (10)$$

where $C_{star}(d)$ is the number of stars in the 24 h of d days ago, and $C_{download}(d)$ is the number of downloads in the 24 h of d days ago.

3.2 Method to Get the Target Front-End Frameworks

In this study, we obtained the front-end frameworks implemented in the RealWorld project [7], which use JavaScript as the programming language. Among them, we eliminated patterns such as "using the same framework, but only the bundling tools are different". As a result, as of October 2021, we obtained the following 17 frameworks: Angular, Apprun, Aurelia, Dojo, Ember.js, Hyperapp, Imba, Neo, Owl, Preact, React, Riot, San, Solid, Stencil, Svelte, and Vue.

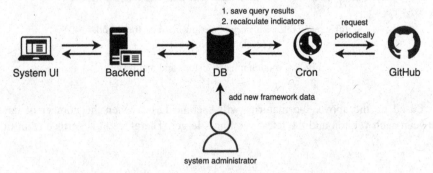

Fig. 3. Schematic architecture of the proposed system

3.3 Implementation of the Proposed Selection Assistance System

Figure 3 shows the schematic architecture of the proposed system. The system sends a request to GitHub periodically by Cron program to get new data and save it in the database. In addition, the system recalculates the indicators and saves the calculation results in the database. The calculation results are displayed in the UI of the system via the Backend program. In the future, when new frameworks are released, the system operator adds the name of the framework, repository, public name in npm, and link to the RealWorld project to the framework table in the database. By doing so, the system is designed to automatically retrieve the data of the new framework and calculate the indicators when the Cron program is executed next time.

3.4 UI of the Proposed System

Figure 4 shows the UI of the proposed system. It is provided as a web application and displays the indicator values and weighted scores of each framework in a table. At the bottom of the table, there is an input field to set the weight for each indicator in the range of zero to one. By setting these weights, users can specify which indicators they consider important and by how much. The "Weighted Score" column at the right end of the table calculates the value of each framework in real time based on the weights entered by the user. By clicking on each column, the user can sort the table in ascending or descending order. The "Frameworks for Similarity Comparison" section is a function to measure the similarity of APIs between frameworks. By selecting frameworks to be compared in this section, a column will be added to the table and a weight input field is added below it. The following use cases are assumed as selecting comparison targets.

- Select frameworks that the user has already learned, and search for one with the lowest learning cost.
- Select frameworks that has already been used in production and search for one with the lowest migration cost.

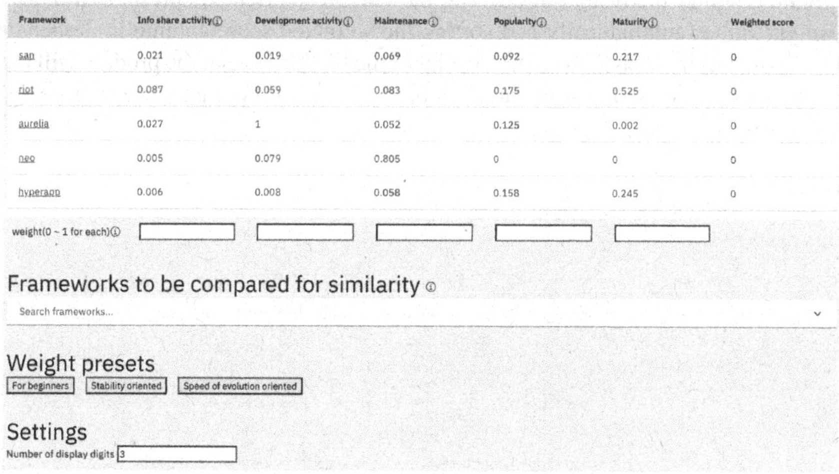

Framework	Info share activity	Development activity	Maintenance	Popularity	Maturity	Weighted score
san	0.021	0.019	0.069	0.092	0.217	0
riot	0.087	0.059	0.083	0.175	0.525	0
aurelia	0.027	1	0.052	0.125	0.002	0
neo	0.005	0.079	0.805	0	0	0
hyperapp	0.006	0.008	0.058	0.158	0.245	0

weight(0 ~ 1 for each)

Frameworks to be compared for similarity

Search frameworks...

Weight presets

| For beginners | Stability oriented | Speed of evolution oriented |

Settings

Number of display digits 3

Fig. 4. UI of the proposed system

Figure 5 shows the system user select "react" as a comparison target. A column "Similarity to react" is added to the table, and the similarity between each framework and react is displayed. Clicking on the buttons in the "Preset weights" section will automatically set the weights that the system prepared in advance. For example, clicking on the "For beginners" button will set the weights of the indicators such as information sharing activity and popularity to a higher value. This saves the user the trouble of specifying the weights in detail.

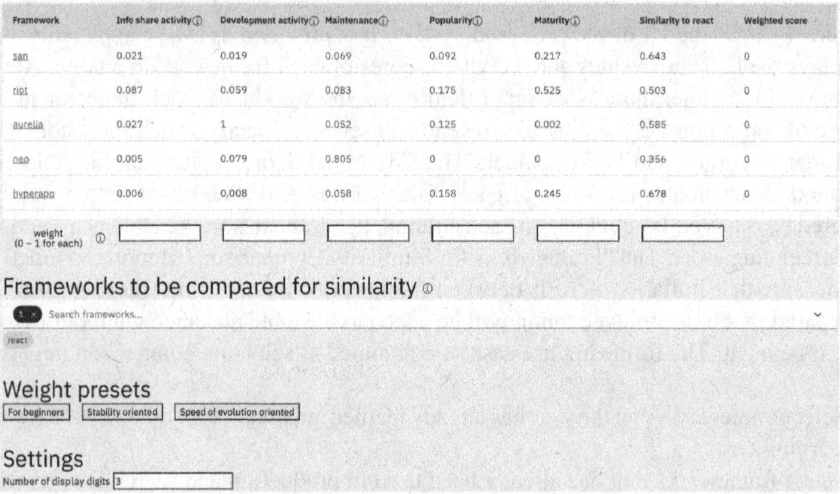

Framework	Info share activity	Development activity	Maintenance	Popularity	Maturity	Similarity to react	Weighted score
san	0.021	0.019	0.069	0.092	0.217	0.643	0
riot	0.087	0.059	0.083	0.175	0.525	0.503	0
aurelia	0.027	1	0.052	0.125	0.002	0.585	0
neo	0.005	0.079	0.805	0	0	0.356	0
hyperapp	0.006	0.008	0.058	0.158	0.245	0.678	0
weight (0 ~ 1 for each) ⓘ							

Frameworks to be compared for similarity ⓘ

1 ✕   Search frameworks...   ⌄

react

Weight presets

[For beginners] [Stability oriented] [Speed of evolution oriented]

Settings

Number of display digits 3

Fig. 5. Selected a target for similarity comparison

Figure 6 shows the results of pushing the "Stability-oriented" preset button to set the weight values automatically and sorting the weighted score column in descending order. The higher the weighted score, the higher the likelihood that the product will meet the user's requirements. The user can click on the name of the framework to access the official documentation of that framework.

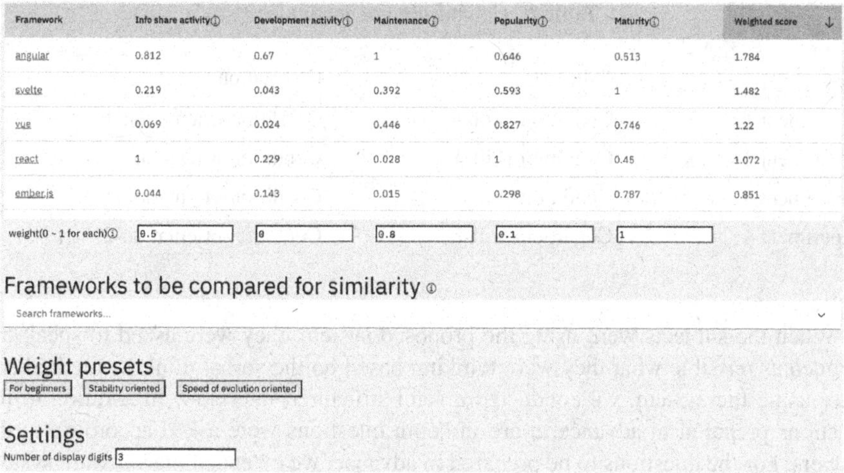

Framework	Info share activity ⓘ	Development activity ⓘ	Maintenance ⓘ	Popularity ⓘ	Maturity ⓘ	Weighted score ↓
angular	0.812	0.67	1	0.646	0.513	1.784
svelte	0.219	0.043	0.392	0.593	1	1.482
vue	0.069	0.024	0.446	0.827	0.746	1.22
react	1	0.229	0.028	1	0.45	1.072
ember.js	0.044	0.143	0.015	0.298	0.787	0.851
weight(0 ~ 1 for each) ⓘ	0.5	0	0.8	0.1	1	

Frameworks to be compared for similarity ⓘ

Search frameworks...   ⌄

Weight presets

For beginners   Stability oriented   Speed of evolution oriented

Settings

Number of display digits 3

Fig. 6. Sorted by weighted score

4 Evaluation Experiment

4.1 Overview of the Experiment

The proposed system was evaluated by conducting semi-structured interviews after it was used by several developers who have experience in front-end development. The scenario of using the system is the following.

When a team of five developers develops a new blog application as specified in the RealWorld [7] specification, the subject selects one framework to use. In this case, the subject uses the proposed system to narrow down the candidates to three.

In addition, two types of conditions, namely skills and orientation, were set for the development members as follows.

Members' skill

Condition-skill-1: the subject's skill is 1, and the skills of other four members are 0.4, 0.3, 0.2, and 0.1, respectively.

Condition-skill-2: the subject's skill is 1, and the skills of other four members are 0.9, 0.8, 0.7, and 0.6, respectively.

Members' orientation

Condition-orientation-1: members are conservative (they think it is better to be as close as possible to the framework already learned)

Condition-orientation-2: members are innovative (they think it is good even if it differs from the framework already learned)

Then, four experiments were conducted by combining the skill and orientation conditions of the development members as in Table 2.

Table 2. Combination of conditions

	Skill	Orientation
Experiment 1	Condition-skill-1	Condition-orientation-1
Experiment 2	Condition-skill-1	Condition-orientation-2
Experiment 3	Condition-skill-2	Condition-orientation-1
Experiment 4	Condition-skill-2	Condition-orientation-2

When the subjects were using the proposed system, they were asked to speak out as much as possible what they were thinking based on the verbal thinking method [9]. After using the system, we conducted a semi-structured interview. In addition to the questions prepared in advance, more in-depth questions were asked according to the answers. For the questions to be prepared in advance, we extended the software system evaluation of Inagaki et al. [10] and referred to the quality characteristics of SQuaRE [11], an international standard for software quality. The questions were broadly divided into the following categories in terms of effectiveness, usability, accuracy of data, and practicality.

Effectiveness

- Did this system effectively compare frameworks?
- Were the six indicators effective in assessing the framework?

Usability

- Is the system's user interface (weight input, preset buttons, etc.) easy to use?
- Was it easy to understand how to use the entire system?

Accuracy of data

- Are the values of the indicators for each framework appropriate?

Practicality

- Can this system be used in the actual selection process?

Table 3 shows the attributes of four subjects for the evaluation experiment.

4.2 Results and Discussions

4.2.1 Improvements to the Algorithm of the Indicators

Interviewer: "Are you satisfied with the evaluation values of each framework?"

Subject 1: "I'm concerned that React's maintenance value is as low as 0.028."

Table 3. Attributes of the four subjects

Subject	Years of programming	Years of developing front-end	The best front-end frameworks	Front-end frameworks the subject knows
1	4	2	React	Vue, Angular, Svelte
2	11	4	Vue	React, Svelte, Angular, Riot, (Backbone, Knockout) [*]
3	5	2	Svelte	React, Svelte, Angular
4	6	4	React	Vue, Angular, Svelte, Solid, Riot

[*]frameworks in parentheses mean that the proposed system does not have their data.

The reason is that React is the most popular frameworks, but it has a large number of issues, thus many issues are left unresolved. In other words, the sub-score of the degree of neglecting the issues in the maintenance activity becomes larger, and the maintenance activity score becomes smaller, which is expressed in Eq. (8). Among the React issues, we found that some of the neglected issues were caused by incomplete bug reports (e.g., no reproduction procedure). This is not a problem on the part of the framework developers, so we can exclude them to prevent the degree of neglecting the issues from becoming too large.

Interviewer: "Are there any problems that you have found after using this system?"

Subject 1: "Famous frameworks are more likely to be near the top. Frameworks that are not well known but have good performance will be truncated."

The reason is that the indicators for the information sharing activity and maintenance activity are calculated by adding up the data. Specifically, if the framework is famous, the number of GitHub issues and Stack Overflow questions is likely to increase, and the score is likely to become larger when these data are added up. As a suggestion for improvement, the first thing to do is to calculate the score as a percentage instead of an additive method. For example, in the case of active information sharing, the score could be calculated based on the percentage of questions that are answered. In addition to the six indicators, increasing the number of indicators that are not related to fame, such as performance, could be a solution to this problem.

4.2.2 Improvements to the System's User Interface

In the proposed system, as shown in Fig. 7, the definitions do not pop up until the mouse is hovered over the icon in each column. Subjects 1 and 3 checked all the definitions in detail, while Subjects 2 and 4 only checked some of the indicators. If the definitions are not checked, the subjects' perception of the indicators may be misaligned, and the calculation results may be different from the subjects' true orientation. Therefore, a possible improvement would display the definition in a pop-up even when the weights are being focused and inputted.

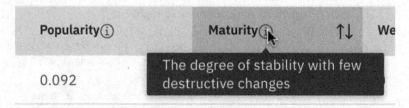

Fig. 7. Definition popping up in the system

4.2.3 Effectiveness of the System

When asked if the system was able to compare frameworks effectively, all four subjects responded positively to the method of inputting weights for the indicators and comparing the scores. Subjects 2 and 3 appreciated the fact that they could deal with different skills and orientations of the development team members by changing the weight values.

> Interviewer: "Is this system likely to be used in the actual selection process?"
>
> Subject 1: "If more indicators are added, it will be more practical. It seems to have enough utility even at now."
>
> Subject 2: "Looks like a good first step to narrow down frameworks. Especially helpful for those who are not familiar with the front-end."
>
> Subject 3: "Since the weight value can be set freely, it seems to be very versatile. It may be useful for various development teams."
>
> Subject 4: "It seems to have a practical feature. But it would be nice if there was a simple tutorial on how to use it, because it seems difficult to use it without an explanation for the first time."

All four subjects answered that the proposed system has some practicality. From the answer of Subject 2, it is expected that the proposed system is useful especially for developers who are not familiar with the front-end. It is also expected from the answer of Subject 3 that it is versatile enough to be used by various development teams. Based on the responses of Subjects 1 and 4, the proposed system needs to add the following two functions to improve its practicality: more indicators, and tutorial on how to use the system.

5 Conclusion

In this study, to reduce the difficulties in selecting a front-end framework, we have developed a selection assistance system that collects and analyzes data and presents the result to developers. This system uses six quantitative indicators, Development activity, Information sharing activity, Similarity of APIs with a specific framework, Maintenance activity, Maturity, and Popularity. To deal with differences in the skills and orientations of the various development teams, our system allows users to specify the importance of an indicator by adjusting the weight. The system has been designed to be compatible not only with existing frameworks, but also with future ones. Furthermore, we conducted evaluation experiments with several front-end developers on four practical scenarios and found that the method of allowing users to set the weights of each indicator is effective and practical.

The code for the proposed system has been made public in GitHub repository: https://github.com/KoichiKiyokawa/tech-choice/. And the URL for the proposed system is https://tech-choice.web.app/.

Future work includes further investigation and improvement of definition and algorithm for each indicator and adding new indicators, such as performance.

References

1. Larios Vargas, E., Aniche, M., Treude, C., Bruntink, M., Gousios, G.: Selecting third-party libraries: the practitioners' perspective. In: Proceedings of the 28th ACM Joint Meeting on European Software Engineering Conference and Symposium on the Foundations of Software Engineering, pp. 245–256. Association for Computing Machinery (2020)
2. Gizas, A., Christodoulou, S., Papatheodorou, T.: Comparative evaluation of JavaScript frameworks. In: Proceedings of the 21st International Conference on World Wide Web, WWW 2012 Companion, New York, NY, USA, pp. 513–514. Association for Computing Machinery (2012)
3. Graziotin, D., Abrahamsson, P.: Making sense out of a jungle of JavaScript frameworks. In: Heidrich, J., Oivo, M., Jedlitschka, A., Baldassarre, M.T. (eds.) PROFES 2013. LNCS, vol. 7983, pp. 334–337. Springer, Heidelberg (2013). https://doi.org/10.1007/978-3-642-39259-7_28
4. Kaluža, M., Troskot, K., Vukelić, B.: Comparison of front-end frameworks for web applications development. In: Zbornik Veleučilišta u Rijeci, Vol. 6, No. 1, pp. 261–282 (2018)
5. Abdellatif, A., Zeng, Y., Elshafei, M., Shihab, E., Shang, W.: Simplifying the search of npm packages. Inf. Softw. Technol. **126**, 106365–106376 (2020)
6. Bissyandé, T.F., Thung, F., Lo, D., Jiang, L., Réveillére, L.: Orion: a software project search engine with integrated diverse software artifacts. In: Proceedings of 2013 18th international conference on engineering of complex computer systems, IEEE. pp. 242–245 (2013)
7. Sun, Y.: A real-world SPA. In: Practical application development with AppRun, pp. 219–246. Apress, Berkeley, CA (2019). https://doi.org/10.1007/978-1-4842-4069-4_10
8. Hilton, M., Nelson, N., Tunnell, T., Marinov, D., Dig, D.: Trade-offs in continuous integration: assurance, security, and flexibility. In: Proceedings of the 2017 11th Joint Meeting on Foundations of Software Engineering, New York, NY, USA, Association for Computing Machinery, pp. 197–207 (2017)
9. Ericsson, K.A., Simon, H.A.: Protocol Analysis: Verbal Reports as Data. The MIT Press (1984)
10. Inagaki, S., Funaoi, H., Yamaguchi, E.: Development and evaluation of reconstructive concept mapping software. J. Sci. Educ. Japan **25**(5), 304–315 (2001). (in Japanese)
11. Komiyama, S., Azuma, M.: Establishment of international standards for system and software quality: Japan's initiative for international standardization. Digit. Pract. J. Inf. Process. Japan **10**(1), 62–73 (2019). (in Japanese)

Data Science: Architectures

Autonomous Real-Time Science-Driven Follow-up of Survey Transients

Niharika Sravan[1](✉), Matthew J. Graham[1], Christoffer Fremling[1], and Michael W. Coughlin[2]

[1] Division of Physics, Mathematics, and Astronomy,
California Institute of Technology, Pasadena, CA 91125, USA
[2] School of Physics and Astronomy, University of Minnesota,
Minneapolis, MN 55455, USA

Abstract. Astronomical surveys continue to provide unprecedented insights into the time-variable Universe and will remain the source of groundbreaking discoveries for years to come. However, their data throughput has overwhelmed the ability to manually synthesize alerts for devising and coordinating necessary follow-up with limited resources. The advent of Rubin Observatory, with alert volumes an order of magnitude higher at otherwise sparse cadence, presents an urgent need to overhaul existing human-centered protocols in favor of machine-directed infrastructure for conducting science inference and optimally planning expensive follow-up observations.

We present the first implementation of autonomous real-time science-driven follow-up using value iteration to perform sequential experiment design. We demonstrate it for strategizing photometric augmentation of Zwicky Transient Facility Type Ia supernova light-curves given the goal of minimizing SALT2 parameter uncertainties. We find a median improvement of 2–6% for SALT2 parameters and 3–11% for photometric redshift with 2–7 additional data points in g, r and/or i compared to random augmentation. The augmentations are automatically strategized to complete gaps and for resolving phases with high constraining power (e.g. around peaks). We suggest that such a technique can deliver higher impact during the era of Rubin Observatory for precision cosmology at high redshift and can serve as the foundation for the development of general-purpose resource allocation systems.

Keywords: Real-time resource allocation · Astronomical surveys · Transients · Type Ia supernovae

1 Introduction

Over the past two decades, increasingly sophisticated all-sky surveys have ushered in a golden era of time-domain astronomy. Not only has it produced unprecedented statistics [10,23,25] and revealed an astonishing diversity of transients [4,36,38], but has paved the way for putting together a rich understanding

© Springer Nature Switzerland AG 2022
S. Sachdeva et al. (Eds.): BDA 2021, LNCS 13167, pp. 59–72, 2022.
https://doi.org/10.1007/978-3-030-96600-3_5

of the physical processes governing our Universe. More recently, our ability to probe transients synchronously in electromagnetic and gravitational waves have enabled inquiries in regimes previously out of reach [16,30,48]. The advent of the Rubin Observatory in the upcoming decade will transform this paradigm by increasing the discovery rate by at least an order of magnitude over the current rate [31]. However, several limitations prevent us from fully exploiting the science potential of even current survey throughput [20,29,44].

The primary issue is that survey data volumes exceed the capacity of human experts to process. Moreover, in many cases, additional follow-up is necessary to allow detailed science analyses (e.g. obtaining spectra to estimate redshift or multi-wavelength observations to probe circumstellar medium interaction). However, even with a curated list, the number of candidates for follow-up far exceed the availability of suitable resources. For transients, the situation is further exacerbated by the time-sensitivity of decision-making for such allocation while assessing trade-offs between competing science objectives. Unsurprisingly, this situation results in inefficiencies and lost opportunities.

Significant progress has been made in response to these issues. These include data broker systems (e.g., ALeRCE [22]; AMPEL [45]; Antares [52]; Fink [41]; Fritz[1], Lasair [55]; MARS[2]; Pitt-Google Broker[3]), that sort, value-add, and curate alert streams for downstream science groups, and Target and Observation Managers or marshals [32,57], that help them plan and co-ordinate follow-up. In addition, robust and efficient classifiers for complete [9,53], partial [14,40,43], or no light-curves [6] have been developed to aid in follow-up decision-making. Tools to aid in discovery, such as those for identifying out-of-distribution events, have also been developed [35,37,42,58].

An important but relatively underdeveloped component is systems that autonomously strategize optimal follow-up for an arbitrary set of science goals. Designated ORACLEs[4] [56], they share a lot of synergy with brokers and marshals and are an important component of the overall software infrastructure needed to support survey science. Such systems are similar to those for selecting objects to obtain spectra with the goal of overcoming Malmquist bias in training data [5,33,49] except that the prioritization is based on science utility.

While optimal follow-up strategies have been studied for populations of events to determine general guiding principles [12,59], ideally these would be determined on a case-by-case basis and be adaptive to accommodate additional information that becomes available. Recent work [18] solved the optimal resource allocation problem for obtaining galaxy spectra (to derive distances and masses), constrained by an observing budget, to maximize Shannon Information on Ω_m over all measurements. In this work, we solve the problem of optimal photometric follow-up of transient light-curves (LCs) to maximize constraints on theoretical LC models. It differs from [18] in that the design problem is sequential and

---

[1] https://github.com/fritz-marshal/fritz.
[2] https://mars.lco.global/.
[3] https://pitt-broker.readthedocs.io/en/latest/.
[4] Object Recommender for Augmentation and Coordinating Liaison Engine.

executed in real-time. We are motivated to explore photometric augmentation by Rubin Observatory's Legacy Survey of Space and Time (LSST), ~90% of which will consist of the Wide Fast Deep (WFD) survey with a nominal cadence of ~3 days in any filter (u, g, r, i, z, or y) and ~15 days in the same filter. It is unclear to what extent the resulting sparse LCs could be directly interpreted with theoretical models. This is especially true for fast-evolving transients, including many core-collapse supernovae (SNe). Value-driven augmentation would maximize science uniquely possible due to LSST especially at the fainter end.

In this paper, we focus on optimally augmenting SN Ia LCs from Phase-I of the Zwicky Transient Facility (ZTF) [7,24] public survey for the goal of maximizing constraints on SALT2 model parameters [8]. However, the framework is generalizable to any transient type and set of theoretical models. In Sect. 2 we present the problem statement and our solution for strategizing real-time follow-up under uncertainty. We present results from our method in Sect. 3 and conclude in Sect. 4.

We assume a flat ΛCDM cosmology with $H_0 = 70\,\mathrm{km/s/Mpc}$, $\Omega_m = 1 - \Omega_\Lambda = 0.3$.

2   Problem Statement and Algorithms

We aim to augment ZTF Phase-I branch-normal SN Ia LCs in g and r with photometry in g, r, and i using the same instrument to maximize constraints on SALT2 models [26]. SALT2 models are used widely in cosmological analyses to measure distances using the Phillips' relation [47]. They are described by three parameters, x_0, x_1, and c, where x_0 is a scaling parameter, x_1 is a stretch parameter, c is a color term. During Phase-I, ZTF conducted a public survey of the visible Northern sky every 3 days and the visible Galactic plane every day in g and r using a ~47 deg^2 imager on the 48-in. telescope on Palomar. We include i-band augmentation in our analysis to ascertain the importance of follow-up in different regimes. As such, data in redder filters are also important for deriving physics, including understanding the explosion mechanism [21] and estimating H_0 [11]. To ensure we can accurately model the LC in i-band we select SNe Ia for which there is at least one i-band data within 60 days of trigger. SNe Ia used here were classified by the Bright Transient Survey (BTS), which aims to conduct a complete census of $r \lesssim 18.5$ mag transients [23,46] using a low resolution IFUS on the Palomar 60-in. telescope [51].

2.1   Optimal Real-Time Decision Under Uncertainty

The problem of strategizing optimal follow-up in real-time can be thought of as follows. We are sequentially presented with available actions to choose from ($\mathcal{A}$, e.g. observe in a certain passband), each associated with some cost (e.g. exposure time), that map to a distribution of outcome states ($\mathcal{S}$, dictated by nature e.g. observed magnitude) and some utility (dictated by our objective, e.g. improvement in model constraints). In other words, we must assess the explore-exploit tradeoff between available actions given their expected utility, we cannot undo past actions but should adapt to new information, and there is some uncertainty about the future. Moreover, we may expect some information to become available in the future (e.g. if a survey is scheduled to revisit a field in some interval). The goal is to find the optimal set of actions, constrained by a budget, that maximizes the net utility of outcome states.

The solution is to choose the action with the maximum expected utility at every timestep, where the expected utility of action $\mathcal{A}$, $EU(\mathcal{A})$, is given by:

$$EU(\mathcal{A}) = \int_{\mathcal{S}} dP_{\mathcal{A}}(\mathcal{S}) U(\mathcal{S}) \tag{1}$$

where $P_{\mathcal{A}}(\mathcal{S})$ is the probability of outcome state $\mathcal{S}$ for an action $\mathcal{A}$ and $U(\mathcal{S})$ is the utility of state $\mathcal{S}$. The choice of utility conveys our preference for a given outcome state. Popular choices are A- or D-optimal designs, that minimize the average variance or maximize the Shannon Information of model parameters, respectively. However, more sophisticated utility functions, to aid in model discrimination [12], min-maxing, among others, also exist. In this work we use a proxy for A-optimality as our choice for utility (see Sect. 2.2). We assume a discrete action space (one visit per night in a given filter), a cost of one per action, and a fixed budget per SN Ia.

In practice, it can be computationally difficult to evaluate $EU(\mathcal{A})$ for all actions at a given timestep (i.e. all possible combinations of follow-up at the current and all future epochs). It can also be difficult to compute the utility (e.g. computing the Information matrix using sampling), especially for continuous outcome states. It is therefore useful to bound the problem and/or approximate the utility function, as in a deep Q-network [39]. We discuss our simplifications for the problem of augmenting ZTF SN Ia LCs to minimize SALT2 uncertainties next.

2.2   Sequential 'Pseudo A-Optimal' Experiment Design

First, instead of finding the optimal action set for the full budget across the entire episode, we find the optimal set at the current epoch, i.e.

$$\mathcal{A} = \{\emptyset, g, r, i, gr, ri, ig, gri\} \tag{2}$$

and allocate the remaining budget randomly across future epochs. This substitutes future optimal actions for future mean actions. Second, we assume a fixed outcome state per action, i.e.

$$dP_{\mathcal{A}}(\mathcal{S}) = \delta(S_{\mathcal{A}}^* - S) d_{\mathcal{A}}(\mathcal{S}) \tag{3}$$

where $\mathcal{S}^*$ is outcome state associated with $\mathcal{A}$, i.e. $S_{\mathcal{A}}^* = \{$photometry$(\mathcal{A})$, remainingbudgetrandom, observedphotometry, futuresurveyphotometry$\}$ and δ is the Kronecker delta function. We adopt the inverse mean SALT2 parameter uncertainty as our utility for $\mathcal{S}$. This is meant to serve as a proxy for A-optimality, since sncosmo solves the χ^2 minimization problem, which is not equal to a maximum likelihood estimate. Fitting SALT2 models requires an estimate of the source redshift since the LCs are fit in rest-frame. Since this may not be available in real-time settings, we solve for the photometric redshift along with SALT2 parameters. We do, however, apply Milky-way extinction [13,54] given the known SN sky location. Since, in many cases, $EU(\mathcal{A})$ differ numerically by only a very small value, we require an improvement of ϵ over the expected utility of no action to take that action. A smaller value of ϵ results in greedy actions and spending the budget early. We tune the hyperparameter ϵ given a budget to maximize $EU(\mathcal{A})$ for all ZTF SNe Ia.

Every night, we estimate the $EU(\mathcal{A})$ using Eqs. 1 and 3. This requires estimating $S_{\mathcal{A}}^{*}$. In real-time applications, this can be estimated using machine-learning based forecasting. In this paper, we focus on bracketing the upper limit performance of our framework by using a hypothetical forecaster (we report on real-time performance using an encoder-decoder LSTM in an upcoming work; Sravan et al., in prep). We use 2-D Gaussian Process regression on the full observed LC in g, r, and i (as discussed in [56] with minor modifications to operate in magnitude space) as our hypothetical forecaster. Estimates using this yield a realistic but upper limit performance since it leverages future photometry to estimate the outcome for the present state. We do not use SALT2 or any SN Ia LC models based on the Phillips relation [47] for this purpose because our science objective is intended to maximize constraints on the same and doing so would feed back any model errors or biases [34,50]. Given mean magnitudes from our Gaussian Process regressor, we simulate magnitude uncertainties using a skew normal fit to BTS SN magnitude uncertainties in each passband and magnitude bin[5]. We do not account for magnitude uncertainty correlations on a given night.

Estimating $S_{\mathcal{A}}^{*}$ also requires estimating future survey behaviour. For this we simulate an observing strategy for ZTF Phase-I in g and r by drawing revisit cadences from a Kernel Density Estimate (KDE)[6] for revisit intervals for BTS SNe. We do not account for gaps in cadence due to the moon or systematic difference in cadence for brighter/fainter SNe. For every revisit we assign a random night's observing strategy in a random BTS SNe. This assumes independence of observing strategies per visit which we verify using a χ^{2}-contingency test. Our test yielded a p-value of 0.92 and we fail to accept our null hypothesis, that two consecutive nights' observing strategy, defined as a sequence of observed passbands, are dependent. We then estimate photometry at the simulated observing strategy using the method described above.

Finally, we take the action with the maximum expected utility and repeat the next day. To ensure robustness to stochastic outcomes for future survey sampling (described above), we simulate the $S_{\mathcal{A}}^{*}$ ten times and take the modal action with maximum $EU(\mathcal{A})$ across all simulations. We do not augment if ZTF observed that day from the true observed LC to avoid duplicate observations. To simulate

---

[5] Bin edges at 12, 15.5, 17.5, 19.5, and 21.5 mag.
[6] Gaussian kernel with bandwidth of 0.005.

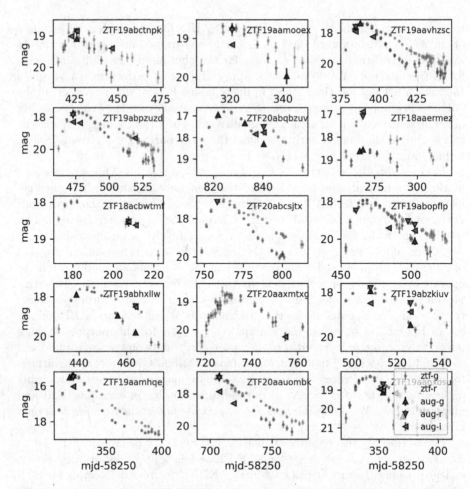

Fig. 1. Random sample of ZTF SN Ia LCs augmented in real-time given the goal of minimizing SALT2 parameter and photometric redshift uncertainties. Translucent green and red error-bars with circular markers are ZTF difference imaging photometry in g and r, respectively. Green (upright), red (downwards), and yellow (left-pointing) triangular markers are simulated augmented photometry in g, r, and i. The augmentations are typically strategized around first/second peaks and valleys and to complete gaps in LCs. (Color figure online)

observed photometry given a chosen action, we use the method described earlier for estimating $S_{\mathcal{A}}^*$. The simulated observed photometry are made available to determine the observing strategy for the next day.

3   Science Gain from Real-Time Photometric Augmentation of ZTF SNe Ia

Table 1. Median (IQR) improvement for SALT2 parameter (x_0, x_1, and c) and photometric redshift (z) uncertainty over random augmentation of SNe Ia LCs in g, r, and i.

Budget	$\delta\sigma_{x_0}$	$\delta\sigma_{x_1}$	$\delta\sigma_c$	$\delta\sigma_z$	ϵ	Expense	Frac Aug
3	0.02 (0.21)	0.03 (0.18)	0.05 (0.21)	0.06 (0.51)	0.78	2	0.34
6	0.03 (0.24)	0.05 (0.22)	0.04 (0.22)	0.06 (0.71)	0.92	5	0.11
9	0.05 (0.24)	0.06 (0.20)	0.05 (0.22)	0.11 (0.64)	0.96	7	0.08

Notes -
Expense: Median utilization of budget
Frac aug: Fraction of events that received at least one augmentation

Table 1 shows the improvement for SALT2 parameter and photometric redshift uncertainty over random augmentation with the same expenditure. Note that not entire the budget is exhausted for all SNe Ia. However, of 497 SNe Ia ∼70–90% received at least one augmentation. As expected, ϵ decreases with budget. This is because the threshold of taking an action is lower if more budget is available. Figure 1 shows example ZTF SNe Ia augmented using our framework given a budget of six. Typically, augmentations are strategized to fill gaps, and around LC peaks and valleys. While we do not assume a spectroscopic redshift while strategizing follow-up, if spectroscopic redshift is eventually used (using either the SN or host spectrum) to fit the LCs the improvement decreases by ∼2%. While the associated mean values for x_0 and z are not affected much, those for x_1 and c change by 4–8% and 8%, respectively. We leave the exploration of the impact on cosmological analyses to upcoming work.

Figure 2 shows the number of augmentations vs sparsity of the ZTF survey LCs given a budget of six. As mentioned earlier, sparse LCs result more augmentations. Sparser LCs also experience more improvement. For instance, events with less than an average of one photometry every three days points (similar to that of LSST WFD) experience 2–5% more improvement than random allocation (7%, 7&, 9%, and 10% for σ_{x_0}, σ_{x_1}, σ_c, and σ_z for a budget of 6). This suggests greater returns from such an approach for LSST-like LCs. Figure 3 shows the distribution of the number of augmented photometry. Most augmentations are strategized in the i-band followed by g, then r. This is because the second peak of a SN Ia LC is most pronounced in the i-band. On the other hand, even though the second peak is more prominent in r than g, since g-band LCs are typically fainter than r, it has sparser coverage and thus receives more augmentations.

Figure 4 shows the distribution of phase at which augmentations are strategized. Here t_0 is time of the bolometric LC peak from `sncosmo` fits to ZTF public survey LCs. Half the augmentations are solicited within 2 days of peak

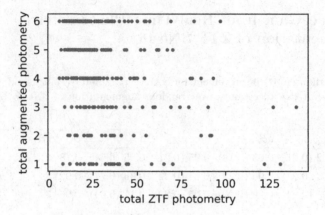

Fig. 2. Total number of augmented vs ZTF public survey photometry. Augmentations are strategized to fill in sparse LCs.

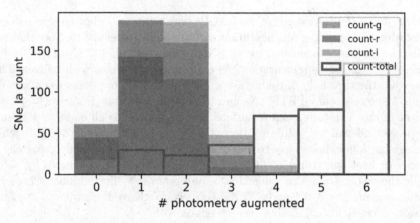

Fig. 3. Distribution of number of photometry augmented. There are slightly more augmentations in *i*, than *g*, followed by *r*.

and 90% of are strategized within 26 days. The distributions have three modes that become more pronounced at redder wavelengths. This is once again because augmentations are strategized around LC phase with high variability, i.e. around peaks and valleys.

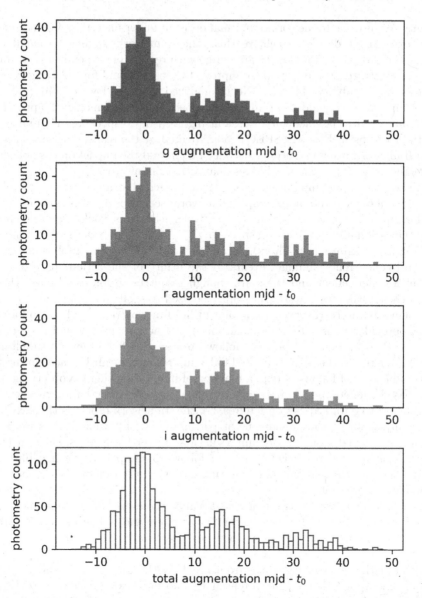

Fig. 4. Distribution of time to/since peak (t_0) at which photometry is augmented (budget = 6). Augmentations are typically strategized around first and second peaks/valleys.

4 Conclusions

In this paper, we solve the problem of real-time optimal resource allocation and demonstrate it for the case of photometric follow-up of ZTF SNe Ia given the goal of minimizing mean SALT2 parameter and photometric redshift uncertainty. We use the framework of optimal decision under uncertainty and sequential experiment design to achieve this. The framework produces a small but robust improvement for target parameters with reasonable expenditure. It automatically strategizes follow-up to improve LC sparsity and for resolving phases with high variability. Though SNe Ia analyzed here are low redshift ($\lesssim 0.15$), such an approach could yield important constraints for cosmology if applied to photometrically classified survey LCs at high redshift, especially in conjunction with programs such as RESSPECT [33]. We analyze the realistic performance of our framework using an encoder-decoder LSTM and the impact on cosmological inference, e.g. H_0 or the Hubble residual, in upcoming work.

There are a few modifications to our approach that are necessary to allow tackling of diverse and more complex real-world scenarios. First is expanding the follow-up action space to include more wavelengths. This would require increasing the cardinality of the action space to $\sum_{i=0}^{N} \binom{N}{i}$, where N is the total number of follow-up regimes (e.g. filters). Second is strategizing optimal follow-up given several concurrent candidate events. If the utility of the action space of each event is independent and there are enough resources to follow-up all of them, then the existing framework can simply be used to strategize follow-up for each of them separately. However, for a joint utility function (e.g. if follow-up for one event provides information about another), the action space would need to be increased to evaluate all possible follow-up scenarios for all events simultaneously. Related to the issue of limited follow-up resources, third, is accounting for a variable cost of follow-up (e.g. a function of integration time), a observing season based budget, and any data acquisition latency. We note that our framework is tolerant to data acquisition failures, i.e. the framework proceeds the next day normally, ignoring the previous night's strategy. Finally, while in this work we focus on optimal follow-up for improving parameter constraints, such an approach can be used to design long-term follow-up campaigns, for e.g. building priors on physical parameters or discriminating between theoretical models for populations of events.

In this work, we focus on SNe Ia because these are quite abundant, relatively well-defined, with robust theoretical models and public modeling infrastructure supporting it. However, there are potentially other science use cases that are more suited to this type of method. Typically, these would be for LCs that are either sparse (e.g. those from LSST WFD) and/or fast-evolving. One important application is the potential to inform optimal follow-up of LIGO gravitational-wave kilonova counterparts [2,17], constrain H_0 [1,19,28] or discriminate between binary neutron star and neutron star-neutron star LC models [3,15,27]. Given the high survey data volumes, limited expert time and follow-up resources, delegating autonomous systems to handle data acquisition would free up time to not only further improve the science return for the particular field being targeted, but also for other science cases outside the purview of machine intelligence.

References

1. Abbott, B.P., et al.: A gravitational-wave standard siren measurement of the Hubble constant. Nature **551**(7678), 85–88 (2017). https://doi.org/10.1038/nature24471

2. Ackley, K., et al.: Observational constraints on the optical and near-infrared emission from the neutron star-black hole binary merger candidate S190814bv. A&A **643**, A113 (2020). https://doi.org/10.1051/0004-6361/202037669

3. Anand, S., et al.: Optical follow-up of the neutron star-black hole mergers S200105ae and S200115j. Nat. Astron. **5**, 46–53 (2021). https://doi.org/10.1038/s41550-020-1183-3

4. Arcavi, I.: Rapidly rising transients in the supernova—superluminous supernova gap. ApJ **819**(1), 35 (2016). https://doi.org/10.3847/0004-637X/819/1/35

5. Astudillo, J., Protopapas, P., Pichara, K., Huijse, P.: An information theory approach on deciding spectroscopic follow-ups. AJ **159**(1), 16 (2020). https://doi.org/10.3847/1538-3881/ab557d

6. Baldeschi, A., Miller, A., Stroh, M., Margutti, R., Coppejans, D.L.: Star formation and morphological properties of galaxies in the pan-STARRS 3π survey. I. A machine-learning approach to galaxy and supernova classification. ApJ **902**(1), 60 (2020). https://doi.org/10.3847/1538-4357/abb1c0

7. Bellm, E.C., et al.: The zwicky transient facility: system overview, performance, and first results. PASP **131**(995), 018002 (2019). https://doi.org/10.1088/1538-3873/aaecbe

8. Betoule, M., et al.: Improved cosmological constraints from a joint analysis of the SDSS-II and SNLS supernova samples. A&A **568**, A22 (2014). https://doi.org/10.1051/0004-6361/201423413

9. Boone, K.: Avocado: photometric classification of astronomical transients with Gaussian process augmentation. AJ **158**(6), 257 (2019). https://doi.org/10.3847/1538-3881/ab5182

10. Brout, D., et al.: First cosmology results using Type Ia supernovae from the Dark Energy Survey: photometric pipeline and light-curve data release. ApJ **874**, 106 (2019). https://doi.org/10.3847/1538-4357/ab06c1

11. Burns, C.R., et al.: The Carnegie supernova project: absolute calibration and the Hubble constant. ApJ **869**(1), 56 (2018). https://doi.org/10.3847/1538-4357/aae51c

12. Carbone, D., Corsi, A.: An optimized radio follow-up strategy for stripped-envelope core-collapse supernovae. ApJ **889**(1), 36 (2020). https://doi.org/10.3847/1538-4357/ab6227

13. Cardelli, J.A., Clayton, G.C., Mathis, J.S.: The relationship between infrared, optical, and ultraviolet extinction. ApJ **345**, 245 (1989). https://doi.org/10.1086/167900

14. Carrasco-Davis, R., et al.: Alert classification for the ALeRCE broker system: the real-time stamp classifier. arXiv e-prints arXiv:2008.03309, August 2020

15. Coughlin, M.W., Dietrich, T.: Can a black hole-neutron star merger explain GW170817, AT2017gfo, and GRB170817A? Phys. Rev. D **100**(4), 043011 (2019). https://doi.org/10.1103/PhysRevD.100.043011

16. Coughlin, M.W., Dietrich, T., Margalit, B., Metzger, B.D.: Multimessenger Bayesian parameter inference of a binary neutron star merger. MNRAS **489**(1), L91–L96 (2019). https://doi.org/10.1093/mnrasl/slz133

17. Coughlin, M.W., et al.: GROWTH on S190425z: searching thousands of square degrees to identify an optical or infrared counterpart to a binary neutron star merger with the zwicky transient facility and palomar gattini-IR. ApJ **885**(1), L19 (2019). https://doi.org/10.3847/2041-8213/ab4ad8

18. Cranmer, M., Melchior, P., Nord, B.: Unsupervised resource allocation with graph neural networks. arXiv e-prints arXiv:2106.09761, June 2021

19. Dietrich, T., et al.: Multimessenger constraints on the neutron-star equation of state and the Hubble constant. Science **370**(6523), 1450–1453 (2020). https://doi.org/10.1126/science.abb4317

20. Djorgovski, S.G., et al.: Real-time data mining of massive data streams from synoptic sky surveys. arXiv e-prints arXiv:1601.04385, January 2016

21. Folatelli, G., et al.: The Carnegie supernova project: analysis of the first sample of low-redshift type-Ia supernovae. AJ **139**(1), 120–144 (2010). https://doi.org/10.1088/0004-6256/139/1/120

22. Förster, F., et al.: The automatic learning for the rapid classification of events (ALeRCE) alert broker. AJ **161**(5), 242 (2021). https://doi.org/10.3847/1538-3881/abe9bc

23. Fremling, C., et al.: The zwicky transient facility bright transient survey. I. Spectroscopic classification and the redshift completeness of local galaxy catalogs. ApJ **895**(1), 32 (2020). https://doi.org/10.3847/1538-4357/ab8943

24. Graham, M.J., et al.: The zwicky transient facility: science objectives. PASP **131**(1001), 078001 (2019). https://doi.org/10.1088/1538-3873/ab006c

25. Graur, O., et al.: LOSS revisited. II. The relative rates of different types of supernovae vary between low- and high-mass galaxies. ApJ **837**, 121 (2017). https://doi.org/10.3847/1538-4357/aa5eb7

26. Guy, J., et al.: SALT2: using distant supernovae to improve the use of type Ia supernovae as distance indicators. A&A **466**(1), 11–21 (2007). https://doi.org/10.1051/0004-6361:20066930

27. Hinderer, T., et al.: Distinguishing the nature of comparable-mass neutron star binary systems with multimessenger observations: GW170817 case study. Phys. Rev. D **100**(6), 063021 (2019). https://doi.org/10.1103/PhysRevD.100.063021

28. Hotokezaka, K., et al.: A Hubble constant measurement from superluminal motion of the jet in GW170817. Nat. Astron. **3**, 940–944 (2019). https://doi.org/10.1038/s41550-019-0820-1

29. Huerta, E.A., et al.: Enabling real-time multi-messenger astrophysics discoveries with deep learning. Nat. Rev. Phys. **1**(10), 600–608 (2019). https://doi.org/10.1038/s42254-019-0097-4

30. Huth, S., et al.: Constraining neutron-star matter with microscopic and macroscopic collisions. arXiv e-prints arXiv:2107.06229, July 2021

31. Ivezić, Ž, et al.: LSST: from science drivers to reference design and anticipated data products. ApJ **873**(2), 111 (2019). https://doi.org/10.3847/1538-4357/ab042c

32. Kasliwal, M.M., et al.: The GROWTH marshal: a dynamic science portal for time-domain astronomy. PASP **131**(997), 038003 (2019). https://doi.org/10.1088/1538-3873/aafbc2

33. Kennamer, N., et al.: Active learning with RESSPECT: resource allocation for extragalactic astronomical transients. arXiv e-prints arXiv:2010.05941, October 2020

34. Kim, A.G., et al.: Type Ia supernova Hubble residuals and host-galaxy properties. ApJ **784**(1), 51 (2014). https://doi.org/10.1088/0004-637X/784/1/51

35. Lochner, M., Bassett, B.A.: ASTRONOMALY: personalised active anomaly detection in astronomical data. Astron. Comput. **36**, 100481 (2021). https://doi.org/10.1016/j.ascom.2021.100481

36. Lunnan, R., et al.: Two new calcium-rich gap transients in group and cluster environments. ApJ **836**(1), 60 (2017). https://doi.org/10.3847/1538-4357/836/1/60

37. Malanchev, K.L., et al.: Anomaly detection in the Zwicky Transient Facility DR3. MNRAS **502**(4), 5147–5175 (2021). https://doi.org/10.1093/mnras/stab316

38. Margutti, R., et al.: An embedded X-ray source shines through the aspherical AT 2018cow: revealing the inner workings of the most luminous fast-evolving optical transients. ApJ **872**(1), 18 (2019). https://doi.org/10.3847/1538-4357/aafa01

39. Mnih, V., et al.: Human-level control through deep reinforcement learning. Nature **518**(7540), 529–533 (2015). https://doi.org/10.1038/nature14236

40. Möller, A., de Boissière, T.: SuperNNova: an open-source framework for Bayesian, neural network-based supernova classification. MNRAS **491**(3), 4277–4293 (2020). https://doi.org/10.1093/mnras/stz3312

41. Möller, A., et al.: FINK, a new generation of broker for the LSST community. MNRAS **501**(3), 3272–3288 (2021). https://doi.org/10.1093/mnras/staa3602

42. Muthukrishna, D., Mandel, K.S., Lochner, M., Webb, S., Narayan, G.: Real-time detection of anomalies in large-scale transient surveys. arXiv e-prints arXiv:2111.00036, October 2021

43. Muthukrishna, D., Narayan, G., Mandel, K.S., Biswas, R., Hložek, R.: RAPID: early classification of explosive transients using deep learning. PASP **131**(1005), 118002 (2019). https://doi.org/10.1088/1538-3873/ab1609

44. Narayan, G., et al.: Machine-learning-based brokers for real-time classification of the LSST alert stream. ApJS **236**(1), 9 (2018). https://doi.org/10.3847/1538-4365/aab781

45. Nordin, J., et al.: Transient processing and analysis using AMPEL: alert management, photometry, and evaluation of light curves. A&A **631**, A147 (2019). https://doi.org/10.1051/0004-6361/201935634

46. Perley, D.A., et al.: The zwicky transient facility bright transient survey. II. A public statistical sample for exploring supernova demographics. ApJ **904**(1), 35 (2020). https://doi.org/10.3847/1538-4357/abbd98

47. Phillips, M.M.: The absolute magnitudes of Type IA supernovae. ApJ **413**, L105 (1993). https://doi.org/10.1086/186970

48. Raaijmakers, G., et al.: The challenges ahead for multimessenger analyses of gravitational waves and kilonova: a case study on GW190425. arXiv e-prints arXiv:2102.11569, February 2021

49. Richards, J.W., et al.: Active learning to overcome sample selection bias: application to photometric variable star classification. ApJ **744**(2), 192 (2012). https://doi.org/10.1088/0004-637X/744/2/192

50. Rigault, M., et al.: Evidence of environmental dependencies of type Ia supernovae from the nearby supernova factory indicated by local Hα. A&A **560**, A66 (2013). https://doi.org/10.1051/0004-6361/201322104

51. Rigault, M., et al.: Fully automated integral field spectrograph pipeline for the SEDMachine: pysedm. A&A **627**, A115 (2019). https://doi.org/10.1051/0004-6361/201935344

52. Saha, A., et al.: ANTARES: a prototype transient broker system. In: Observatory Operations: Strategies, Processes, and Systems V. Society of Photo-Optical Instrumentation Engineers (SPIE) Conference Series, vol. 9149, p. 914908, July 2014. https://doi.org/10.1117/12.2056988

53. Sánchez-Sáez, P., et al.: Alert classification for the ALeRCE broker system: the light curve classifier. AJ **161**(3), 141 (2021). https://doi.org/10.3847/1538-3881/abd5c1

54. Schlegel, D.J., Finkbeiner, D.P., Davis, M.: Maps of dust infrared emission for use in estimation of reddening and cosmic microwave background radiation foregrounds. ApJ **500**(2), 525–553 (1998). https://doi.org/10.1086/305772

55. Smith, K.W., et al.: Lasair: the transient alert broker for LSST: UK. Res. Notes Am. Astron. Soc. **3**(1), 26 (2019). https://doi.org/10.3847/2515-5172/ab020f

56. Sravan, N., Milisavljevic, D., Reynolds, J.M., Lentner, G., Linvill, M.: Real-time, value-driven data augmentation in the era of LSST. ApJ **893**(2), 127 (2020). https://doi.org/10.3847/1538-4357/ab8128

57. Street, R.A., Bowman, M., Saunders, E.S., Boroson, T.: General-purpose software for managing astronomical observing programs in the LSST era. In: Software and Cyberinfrastructure for Astronomy V. Society of Photo-Optical Instrumentation Engineers (SPIE) Conference Series, vol. 10707, p. 1070711, July 2018. https://doi.org/10.1117/12.2312293

58. Villar, V.A., et al.: A deep-learning approach for live anomaly detection of extragalactic transients. ApJS **255**(2), 24 (2021). https://doi.org/10.3847/1538-4365/ac0893

59. Williamson, M., Modjaz, M., Bianco, F.B.: Optimal classification and outlier detection for stripped-envelope core-collapse supernovae. ApJ **880**(2), L22 (2019). https://doi.org/10.3847/2041-8213/ab2edb

Deep Learning Application for Reconstruction of Large-Scale Structure of the Universe

Kana Moriwaki[✉][ID]

The University of Tokyo, 7-3-1 Hongo, Bunkyo, Tokyo, Japan
kana.moriwaki@phys.s.u-tokyo.ac.jp

Abstract. In this study, we propose to analyze astronomical data obtained in line intensity mapping (LIM) observations using machine learning. The LIM is an emerging method to measure large-scale intensity fluctuations of spectral lines emitted from galaxies and intergalactic medium. Observing their large-scale distributions enables us to study cosmology and galaxy formation and evolution. One of the problems with the LIM is observational noises and line interloper. We develop a three-dimensional convolutional neural network (CNN) that removes those contaminants and extract designated signals from noisy three-dimensional data obtained by LIM. Our CNN has an architecture that encourages extracting the long-range correlations in the spectral direction. We train them on mock observation data generated by simulation and find that they can extract multiple emission lines appropriately.

Keywords: Cosmology and astrophysics · Large-scale structure of the universe · Signal separation · Three-dimensional convolutional neural network · Generative adversarial network

1 Introduction

The ultimate goals of astrophysics include revealing the details of dark matter and dark energy, and the formation and evolution of celestial objects such as galaxies to understand how the universe has evolved over its entire history. For this, astronomical observations, as well as the theory and simulations, are crucial. Interpreting large amounts of observational data requires the use of high-speed and appropriate data processing methods. Future observations will provide survey data over the largest volume ever, but some of them have a serious problem that signals that we are interested in are mixed with the other signals. In this study, we propose to use a machine learning method to solve this problem and demonstrate its performance.

1.1 Background and Motivation

The uniformity of the cosmic microwave background (CMB) observed in the entire sky, coming from 13.8 billion years ago, tells us that the universe was

© Springer Nature Switzerland AG 2022
S. Sachdeva et al. (Eds.): BDA 2021, LNCS 13167, pp. 73–82, 2022.
https://doi.org/10.1007/978-3-030-96600-3_6

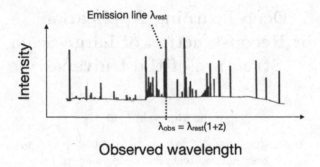

Observed wavelength

Fig. 1. The observed spectrum of a galaxy at redshift z. The emission line with a rest-frame wavelength λ_{rest} is observed at $\lambda_{\mathrm{obs}} = \lambda_{\mathrm{rest}}(1 + z)$.

initially almost uniform right after the Big Bang. Observations with recent ground-based and satellite telescopes have revealed the distribution of galaxies. They are no longer uniformly nor randomly distributed throughout the universe but have characteristic structures, such as filaments and voids, referred to as the large-scale structure of the universe. Cosmological simulations follow the evolution of the universe from the initial conditions motivated by the tiny fluctuations of the CMB to the present universe by solving the gravitational and hydrodynamical equations. Combinations of observations, simulations, and theories have revealed the energy budgets of the universe including the amount of dark energy and dark matter as well as the formation history of the galaxies, but many things still remain unknown. For example, the model that generates the initial condition of the universe, the properties of dark energy and dark matter, and the details of the galaxies in the early universe are not fully understood. To study them, observing distant universe (i.e., the past universe) and larger volume of the universe is crucial.

One of the keys to observing large-scale structures traced by galaxies is to accurately measure the distance of galaxies from us. For this, we can use the emission lines from atoms and molecules in the galaxy, which have certain rest-frame wavelengths λ_{rest}. In the expanding universe, the distance can be measured in terms of redshift, z. It is the shift of the observed wavelength λ_{obs} from the rest-frame wavelength of the emission line, given as

$$\lambda_{\mathrm{obs}} = \lambda_{\mathrm{rest}}(1 + z), \tag{1}$$

Figure 1 shows an example of an observed spectrum of a galaxy.

Traditionally, large-scale structure observations have been done mainly by deeply observing individual galaxies to locate them in three-dimensional space. The largest galaxy surveys (e.g., [8]) have observed galaxy distributions at $z < 1$ and future surveys (e.g., [9]) will be conducted more widely and deeply ever up to $z \sim 2$. While such observations are promising, they generally take a long observational time and it is difficult to survey larger and more distant regions. An emerging method called line intensity mapping (LIM) will play a complementary role to the conventional galaxy surveys. In this observational method, we do not resolve individual galaxies or individual emission lines but measure all the light coming into each pixel. While emission from multiple galaxies could enter a single grid, by measuring the spatial and spectral fluctuations of the intensity, we can detect the underlying large-scale structure of the line emission galaxies as a whole.[1] This allows us to survey larger areas more efficiently than conventional surveys and will provide more constraints on the properties of the universe (see e.g., [16]). Many LIM observations have already started observations [4,5,15] and are planned [6,7,11].

If there is only one emission line to consider, the observed wavelength and redshift (distance) are in one-to-one correspondence (Eq. 1). If there are multiple emission lines, however, there is a problem, and this is the case for most of the wavelength regimes. Let's assume two emission lines, with rest-frame wavelengths λ_1 and λ_2, are emitted from galaxies at redshifts z_1 and z_2, respectively. If they satisfy

$$\lambda_1(1 + z_1) = \lambda_2(1 + z_2), \tag{2}$$

then they are observed at the same wavelength. In the unresolved intensity maps, we cannot separate such contributions, and thus what we observe at each spectral bin is a superposition of the large-scale structures at two different redshifts. This is a so-called line confusion problem. If we use the data confused with such interloper lines, systematic errors would appear in the obtained constraints. In this study, we take a machine learning approach to solve the line confusion problem.

1.2   Related Works

Machine Learning Techniques. In recent years, many astronomical problems have been tackled using machine learning techniques. Convolutional neural network (CNN) plays an important role in the analysis of observational images, in particular, in classification and regression tasks. In astrophysics, we have various generation and reconstruction tasks as well, and they can be dealt with by using generative models such as generative adversarial networks (GANs) [13] and conditional GANs (cGANs) [14]. For instance, GANs have been used to generate mock observational data such as a catalog of galaxies [20] and a large-scale structure traced by neutral hydrogen [24]. cGANs have been used to denoise observational maps such as weak lensing maps [21]. Not only the two-dimensional CNNs but also the application of the three-dimensional CNNs have been studied to analyze observational and simulation data [19,25]

Line De-confusion Methods. Several methods have already been proposed to statistically remove the contributions from the interloper lines. One of them is to compute the cross-power spectrum between the observed intensities and

---

[1] More precisely, it extract the contribution of emission lines by removing the smooth component (continuum emissions) from the observed spectra.

the other tracers of the large-scale structure, such as galaxies, at the redshift of interest [23]. Since the tracers and the interloper signals at different redshifts are not spatially correlated, only the correlation signals from a specific redshift can be extracted if one uses the observational data over a sufficiently large volume. We can also detect statistical signals of fainter lines coming from the distant universe by masking the bright pixels that are considered to be dominated by the foreground interlopers [22]. Another interesting method is to exploit the different metrics at different redshifts. If we adopt a metric defined at a certain redshift, the power spectrum of signals from that redshift would be isotropic in the spectral and angular directions, while signals from the other redshifts would not. The power spectrum of the signals at a certain redshift can then be removed by using this [12].

While statistical signals contain useful information, we can explore many more details if we can isolate the signals at each redshift pixel by pixel. Recently, [3] have proposed a method to reconstruct the intensity distribution at a particular redshift by detecting multiple emission lines from the same galaxy. They demonstrate that their method reproduces the positions of bright sources above given intensity thresholds. They use the spectral data, but not the two-dimensional intensity maps that extend in the angular direction at each wavelength, in which the galaxy correlation information is embedded. In our previous study, we have developed a method to separate multiple signals from such two-dimensional intensity maps using cGANs, and have found that the networks separate two emission lines signals well based on the different degrees of large-scale clustering at different redshifts [18]. In this study, we improve our cGAN method by including the spectral information in the analysis using three-dimensional CNNs.

2   Methods

In this study, we develop cGANs consisting of three-dimensional CNNs for signal separation. The generators take observed data cubes as input and output the line intensity distributions of individual emission lines. In the following, we will consider only two emission lines with rest-frame wavelengths λ_1 and λ_2 for simplicity. The two emission lines from a galaxy at redshift z are observed at $\lambda_1(1 + z)$ and $\lambda_2(1 + z)$, respectively. Therefore the same intensity fluctuation originating from the same large-scale structure appears twice in the spectral direction (see Fig. 2). This long-range correlation in the spectral direction is important for signal extraction. We devise a network architecture that exploits such physical information to make it easy for the networks to learn separation.

We first consider a generator with a simple encoder-decoder architecture and adopt convolution kernels that are elongated along the spectral direction. In particular, we use kernels that are longer than the observed wavelength interval between lines ($> (\lambda_1 - \lambda_2)(1 + z)$) for the first layers of the generator and discriminator. We find that after being trained with mock observational data, the network can extract the designated signals properly even in the presence of

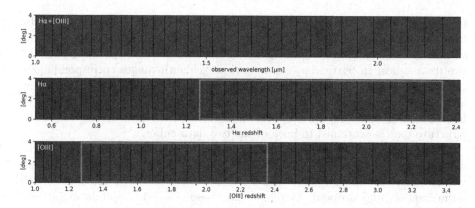

Fig. 2. The intensity fluctuations of the Hα (middle) and [OIII] (bottom) lines and the sum of them (top). The horizontal and vertical axes are the spectral (redshift) direction and angular direction, respectively. The black solid lines are the borders of the spectral bins of SPHEREx observation. The same fluctuation patterns originating from the same galaxy distributions are seen in the Hα and [OIII] intensity (e.g., see those in the yellow boxes enclosing the same redshift range) and they are observed with a $\left(\lambda_{H\alpha} - \lambda_{[OIII]}\right)(1+z)$ separation in the wavelength.

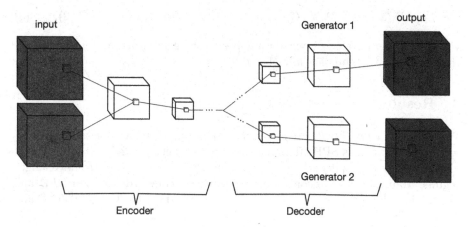

Fig. 3. Generator architecture

observational noise. Interestingly, some of the kernels in the first layers pick up two pixels that are separated by the distance between the wavelengths of the two emission lines. Though this method seems to be promising, we find a flaw in it; it wastes computational costs. To apply this method to observational data with high spectral resolution or to include more than two emission lines, we need to use very long kernels that require more computational resources.

We thus consider a different network architecture in this study. Figure 3 shows an overview of the generator we adopt. The two portions of the observed data cube with different wavelength ranges are input to the generator with one shifted by a certain length in the spectral direction relative to the other. We shift it by the distance between observed wavelengths of the two lines, $(\lambda_1 - \lambda_2)(1 + z)$, so that the signals from the same galaxy come into the corresponding pixels of two data cubes. In particular, we use two wavelength ranges corresponding to the redshift ranges of two yellow boxes in Fig. 2. In the first layer, the generator performs convolution operations on these two data cubes and mixes the information. This allows the networks to more efficiently pick up the large-scale structure that is contained in both of the two data cubes.

The generators have four convolutional layers and four deconvolutional layers which are connected with U-net architecture [14]. The decoder section branches into two and outputs two emission line signals as in Fig. 3. We also adopt batch normalization and dropout. We find that these networks reproduce the true line intensities with better accuracy than those with long convolution kernels discussed above. We note that the network structure does not need to be significantly changed when the spectral resolution changes or different emission lines are added, and thus the computational cost of training does not become huge.

We build conditional Wasserstein GAN (WGAN) [1][2], consisting of two sets of generator and critic, (G_i, D_i) $(i = 1, 2)$ for extracting two emission signals and optimize them with loss functions

$$L_i = D_i(x_i, y_i) - D_i(x_i, G_i(x_1, x_2)) + \lambda_i|y_i - G_i(x_1, x_2)| \quad (i = 1, 2), \qquad (3)$$

where x_1 and x_2 are two input data cubes to the generators, and y_1 and y_2 are the true signals of emission lines 1 and 2.[3]

3   Results

We train the network with mock observational data generated assuming observations with NASA's SPHEREx telescope [10] launched in 2024. It conducts an all-sky survey and a deep survey over 200 deg^2 at a near-infrared wavelength regime. Main target emission lines include those from hydrogen ion (Hα) and oxygen ion ([OIII]) that have the rest-frame wavelengths $\lambda_{H\alpha} = 6563$ Å and $\lambda_{[OIII]} = 5007$ Å, respectively. We generate the training dataset using a halo catalog generation code PINOCCHIO [17] and the emission line model based on [18]. We also add the observational noises at the current expected level in deep regions. For training, we adopt $\lambda_1 = \lambda_2 = 100$. Since the observational data we intend to apply our network to is not yet available, it cannot be evaluated on actual observational data for a while. For this reason, we use mock data for testing. The test dataset is generated with different random seeds than the training dataset.

---

[2] We find that the training becomes unstable when optimizing using vanilla GAN. It is known that we can often avoid such instability by using WGAN.

[3] In practice, the intensities are normalized with an appropriate value. In this study, we normalize them with 1.0×10^{-4} erg/s/cm^2/sr.

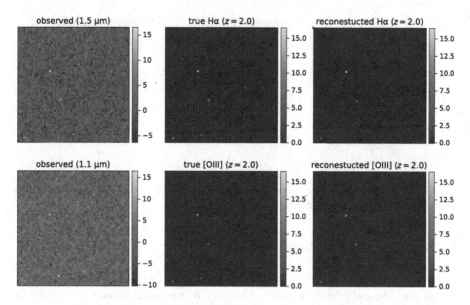

Fig. 4. Reconstruction of Hα and [OIII] line signals.

We find that our network reasonably reconstructs line signals. Figure 4 shows the two-dimensional intensity maps of observed (left), true (middle), and reconstructed (right) data cubes at certain spectral bins, $\lambda_{\mathrm{obs}} = 1.5\,\mu\mathrm{m}$ (top) and $1.1\,\mu\mathrm{m}$ (bottom). The two slices on the left are those of the two input data. Hα and [OIII] lines from the same galaxies at slices of redshift $z = 2$ are observed at these wavelengths. The locations and the intensities of bright emission sources are accurately detected even though the observed images are dominated by large observational noise, especially at $\lambda_{\mathrm{obs}} = 1.1\,\mu\mathrm{m}$.

Figure 5 shows the three-dimensional Hα line intensity maps reconstructed by the network. While the observed data is confused by observational noise and interloper emissions, the network properly reconstructs the emission line signals that trace the underlying large-scale structure. We find that the precisions of the intensity peak detection (i.e., the percentage of true intensity peaks among the reconstructed peaks) are as high as 80% for both Hα and [OIII] intensities. We note that similar detectability is also demonstrated by [3], where they perform spectral analysis for mock observational data of CO emission lines as well as rest-frame optical lines. We also find that the bright ends of the probability distribution functions of the line intensities and the shot noise terms of the power spectra are well reproduced. When the large-scale structure over a large volume is reconstructed with our method, it would be of great importance in studying cosmology as well as galaxy formation and evolution.

Observed   Ground truth   Reconstructed

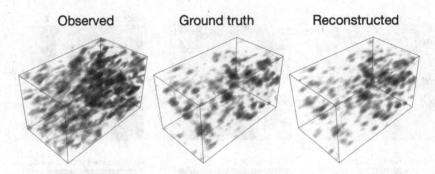

Fig. 5. Reconstruction of the large-scale structure traced by Hα intensity map. The intensity is smoothed for visibility.

4   Summary and Conclusion

The LIM is a unique observational method for observing the large-scale structure over unprecedentedly large volumes. Although its potential impact is great, there is a serious problem of line confusion. So far, many statistical methods have been proposed, but only a few methods have been studied for pixel-by-pixel signal separation. In this study, we propose a method, for the first time, to extract specific emission line signals from the three-dimensional LIM data by analyzing them with three-dimensional CNNs.

We adopt a physics-informed network architecture where the networks are given two observed data cubes at specific wavelengths determined by the rest-frame wavelengths of lines of interest. Using mock observational data generated by simulations, we demonstrate that the line intensity signals are successfully separated with our cGANs. This method can be easily extended to various observations targeting different emission lines.

One of the future tasks is to investigate whether the network trained with mock data will perform as well when applied to actual observational data in the future. There is still uncertainty in the galaxy properties, and our assumptions may differ from those in the real universe. The potential advantage of our method is that the networks can also be trained on spatially correlated noise such as continuum sources and Galactic diffuse emissions. In addition to seeking more realistic line emission models, we can use methods such as blind modeling [2] to build better networks once we have actual observational data.

The large-scale structure reconstructed by our method will provide important information on cosmology and astrophysics. The scale dependence of the power spectrum of the intensity at a specific redshift will provide cosmological constraints, and we can investigate the population of distant galaxies from the probability distribution function of the intensity. In addition, the intensity map itself is important for studying the environmental effects on galaxy formation, and the locations of the clusters may be important for studying the origin of other astronomical signals, including high-energy particles and gravitational waves.

Acknowledgments. The author is supported by JSPS KAKENHI Grant Number 19J21379 and by JSR Felloship.

References

1. Arjovsky, M., Chintala, S., Bottou, L.: Wasserstein GAN. arXiv e-prints arXiv:1701.07875 (2017)
2. Chen, J., Chen, J., Chao, H., Yang, M.: Image blind denoising with generative adversarial network based noise modeling. In: 2018 IEEE/CVF Conference on Computer Vision and Pattern Recognition, pp. 3155–3164 (2018). https://doi.org/10.1109/CVPR.2018.00333
3. Cheng, Y.T., Chang, T.C., Bock, J.J.: Phase-space spectral line de-confusion in intensity mapping. arXiv e-prints arXiv:2005.05341 (2020)
4. Cleary, K.A., et al.: COMAP early science: I. Overview. arXiv e-prints arXiv:2111.05927 (2021)
5. Ade, P., et al.: A wide field-of-view low-resolution spectrometer at APEX: Instrument design and scientific forecast. Astronomy and Astrophysics 642, A60 (2020). https://doi.org/10.1051/0004-6361/202038456
6. Cooray, A., et al.: CDIM: cosmic dawn intensity mapper final report. arXiv e-prints arXiv:1903.03144 (Mar 2019)
7. Crites, A.T., et al.: The TIME-Pilot intensity mapping experiment. In: Holland, W.S., Zmuidzinas, J. (eds.) Millimeter, Submillimeter, and Far-Infrared Detectors and Instrumentation for Astronomy VII. Society of Photo-Optical Instrumentation Engineers (SPIE) Conference Series, vol. 9153, p. 91531W (2014). https://doi.org/10.1117/12.2057207
8. Dawson, K.S., et al.: The baryon oscillation spectroscopic survey of SDSS-III. Astron. J. **145**(1), 10 (2013). https://doi.org/10.1088/0004-6256/145/1/10
9. Aghamousa, A., et al.: The DESI experiment part I: science, targeting, and survey design. arXiv e-prints arXiv:1611.00036 (2016)
10. Doré, O., et al.: Cosmology with the SPHEREX all-sky spectral survey. arXiv e-prints arXiv:1412.4872 (2014)
11. Doré, O., et al.: Science impacts of the SPHEREX all-sky optical to near-infrared spectral survey II: report of a community workshop on the scientific synergies between the SPHEREX survey and other astronomy observatories. arXiv e-prints arXiv:1805.05489 (May 2018)
12. Gong, Y., Silva, M., Cooray, A., Santos, M.G.: Foreground contamination in Lyα intensity mapping during the epoch of reionization. Astrophys. J. **785**(1), 72 (2014). https://doi.org/10.1088/0004-637X/785/1/72
13. Goodfellow, I.J., et al.: Generative adversarial networks. arXiv e-prints arXiv:1406.2661 (2014)
14. Isola, P., Zhu, J., Zhou, T., Efros, A.A.: Image-to-image translation with conditional adversarial networks. CoRR abs/1611.07004 (2016). http://arxiv.org/abs/1611.07004
15. Keating, G.K., Marrone, D.P., Bower, G.C., Keenan, R.P.: An intensity mapping detection of aggregate CO line emission at 3 mm. Astrophys. J. **901**(2), 141 (2020). https://doi.org/10.3847/1538-4357/abb08e
16. Kovetz, E.D., et al.: Line-intensity mapping: 2017 status report. arXiv e-prints p. arXiv:1709.09066 (2017)

17. Monaco, P., et al.: An accurate tool for the fast generation of dark matter halo catalogues. Monthly Notice Royal Astron. Soc. **433**(3), 2389–2402 (2013). https://doi.org/10.1093/mnras/stt907

18. Moriwaki, K., et al.: The distribution and physical properties of high-redshift [O III] emitters in a cosmological hydrodynamics simulation. Monthly Notice Royal Astron. Soc. **481**, L84–L88 (2018). https://doi.org/10.1093/mnrasl/sly167

19. Prelogovic, D., Mesinger, A., Murray, S., Fiameni, G., Gillet, N.: Machine learning astrophysics from 21 cm lightcones: impact of network architectures and signal contamination. Monthly Notice Royal Astron. Soc. (2021). https://doi.org/10.1093/mnras/stab3215

20. Ravanbakhsh, S., Lanusse, F., Mandelbaum, R., Schneider, J., Poczos, B.: Enabling dark energy science with deep generative models of galaxy images. arXiv e-prints arXiv:1609.05796, September 2016

21. Shirasaki, M., Moriwaki, K., Oogi, T., Yoshida, N., Ikeda, S., Nishimichi, T.: Noise reduction for weak lensing mass mapping: an application of generative adversarial networks to Subaru Hyper Suprime-Cam first-year data. Monthly Notice Royal Astron. Soc. **504**(2), 1825–1839 (2021). https://doi.org/10.1093/mnras/stab982

22. Silva, B.M., Zaroubi, S., Kooistra, R., Cooray, A.: Tomographic intensity mapping versus galaxy surveys: observing the Universe in H α emission with new generation instruments. Monthly Notice Royal Astron. Soc. **475**, 1587–1608 (2018). https://doi.org/10.1093/mnras/stx3265

23. Visbal, E., Loeb, A.: Measuring the 3D clustering of undetected galaxies through cross correlation of their cumulative flux fluctuations from multiple spectral lines. J. Cosmol. Astroparticle Phys. **11**, 016 (2010). https://doi.org/10.1088/1475-7516/2010/11/016

24. Zamudio-Fernandez, J., et al.: HIGAN: cosmic neutral hydrogen with generative adversarial networks. arXiv e-prints arXiv:1904.12846 (2019)

25. Zhang, X., et al.: From Dark Matter to Galaxies with Convolutional Networks. arXiv e-prints arXiv:1902.05965 (2019)

Identification of Distinctive Behavior Patterns
of Bots and Human Teams in Soccer

Georgii Mola Bogdan$^{(\boxtimes)}$ [ID] and Maxim Mozgovoy$^{(\boxtimes)}$ [ID]

School of Computer Science and Engineering, The University of Aizu,
Aizuwakamatsu 965-8580, Japan
{d8212101,mozgovoy}@u-aizu.ac.jp

Abstract. The design of human-like AI agents requires evaluation methods that check both robustness of the system and its believability. In this paper we attempt to examine whether is possible to assess similarity of play styles between different human teams and artificial teams in soccer. We rely on "behavior fingerprints" based on heat maps and their comparison using dot product. Our method shows no distinctive differences between the fingerprints of human teams, however, clearly indicates the difference between human teams and artificial teams. This approach is aimed to assist the design of human-like soccer teams but can also be useful in the domain of sports analytics.

Keywords: Soccer · Game AI · Human-likeness · Machine learning

1 Introduction

Being a worldwide popular sport, soccer extended beyond its original scope and became both a serious media business and cultural factor and a de facto testbed for AI research and competitions due to its relatively complex multiagent environment (as exemplified by RoboCup [1]). In addition, soccer-based video games are found among the most popular electronic entertainment products. As any video game it should possess some basic features or "fun factors" that keep the users engaged. One such feature is suspension of disbelief which is considered as indicator of high quality, contributing to player immersion [2]. Suspension of disbelief is usually associated with believability or human-likeness, which is crucial in case of soccer. Since soccer games are played by humans, their artificial representations in a game world are also supposed to be human-like. This task is partially accomplished by using highly detailed models of real athletes and real team structures/symbolics. However, we think that the ultimate contribution to believability can be achieved by interacting with AI agents that possess human-like behavior patterns. Arguably, the most straightforward approach to obtain such agents is to transfer knowledge from real players.

In recent years the role of information technologies in sports and soccer in particular is steadily growing. The development of video tracking systems provides spatiotemporal data that has numerous applications. For example, media companies became able to

S. Sachdeva et al. (Eds.): BDA 2021, LNCS 13167, pp. 83–92, 2022.
https://doi.org/10.1007/978-3-030-96600-3_7

improve spectators' experience, sport analytics experts and coaches can extract information to have a bird's eye view of game tactics or locomotion performance [3]. There also exist works trying to address the problem of performance definition and measurement [4] and game situation modeling [5]. Thus, it may be possible to extract athletes' behavior patterns from real match recordings.

Believability assessment and play style analysis is an integral part of our ongoing research efforts [6, 7]. One of the intermediate goals in this work is to construct an automated assessment approach, which would allow to indicate whether the system under development possesses certain required traits. We are primarily focusing on believability rather than on efficiency due to our goal to provide "believable teams", which is reasonable from game development point of view. While reasonable AI skill level is expected, striving for the best performance is not the primary target for game AI, since it must provide entertainment for players of varying skills. Our current research questions include the following:

- (RQ1) How to design a method that would allow to assess team play styles?
- (RQ2) Does team style persist throughout match and/or between the matches?
- (RQ3) Is it possible to distinguish one team from another?
- (RQ4) Is it possible to distinguish human players from AI-controlled players?

The first question addresses the problem of whether is possible to define properties that represents teams' unique behavior patterns in principle, and whether it is easy to automate their analysis.

The second question is related to the idea that behavior patterns of team in different phases of the game may differ. For example, in the second half of a match physical fatigue of athletes may alter their movements. Next, behavioral patterns of a team may not persist between matches. There are circumstances that may influence team tactics, such as the wish to adapt to the next opponent or to meet some specific subgoal in the ongoing competition. These factors are a part of a "meta-game" and typically not reflected in spatiotemporal datasets.

The third and the fourth questions can be derived form the first two questions: if team behavior styles are identifiable, we should be able to distinguish them. In addition, we should be able to assess "human-likeness" by comparing behavior of human teams with their AI-controlled counterparts. Additional challenges emerge due to scant size of most available spatiotemporal data sets. Even if we obtain all game records of a particular team for the past year or two, the resulting collection would be relatively small for convention machine learning.

The core difficulty in team play style identification lies in a choice of properties representing a unique "team fingerprint". Player and ball tracking datasets provide "low-level" knowledge about team activities, while the logic behind them has to be reconstructed. Experts discuss play style identification, but this work still seems to be in its relatively early stage, mostly limited to discussion of possible options or to in-depth analysis of isolated aspects of the game [8, 9]. Still, the properties analyzed in these works constitute a good starting point.

Since the game of soccer consists mostly of manipulation with a ball and passes, we rely on characteristics of these events as basic features that may reflect individuality.

Possession time of the ball is usually related to the efficiency of a team. In our case, we consider location-based possession of the ball represented with a heatmap. Similarly, we use a heatmap to represent successful pass and receive points. These actions correspond to the moments of the game when the players can do deliberate decision making and thus showing their tactical intentions [4].

2 Datasets

Tracking systems that gather data have been developed by different companies, and their data format is different. That is why preprocessing is necessary to obtain a unified data source. For instance, not all datasets have information about the third (height) coordinate of the ball, or its precision is low. Some sets, such as STATS [10] consist of small anonymized fragments of gameplay. In our work we rely on three independent data sources, obtained from STATS and DataStadium companies, and from Google Research Football environment.

2.1   A. DataStadium: Complete Matches

This dataset is collected and provided by DataStadium Inc. [11]. It consists of five full games played by six Japanese J1 League teams in 2011 season. Some statistical data like team names and formations is also available. The dataset has no event markup, so we had to reconstruct player movements, passes and shots using an automated method described in [12]. We analyze only the data of four teams that played twice to be able to compare team behavior in different matches.

2.2   B. Google Research Football: Virtual Teams

This dataset contain collection of 4800 game sequences recorded during Google Research Football with Manchester City F.C. competition [13]. This dataset contains event markup, which was possible to use after some cleanup.

Google Research's dataset provides a point of reference for the behavior of virtual teams. Unfortunately, in this competition the challenge was to create an AI system controlling the player with the ball (in case of attack) or the player closest to the ball (in case of defense), while the rest of the team is directed by the same built-in rule-based AI engine. Thus, behavioral diversity of virtual teams in the dataset is very limited. Still, the logic of player with the ball has a major impact on the whole team's tactics, so we can start with the presumption that the teams in the dataset are indeed distinct.

For our experiments we took two virtual teams (WeKick and SaltyFish) that played the largest number of matches in the dataset. As result we use 501 game sequences for SaltyFish and 432 sequences for WeKick team (each sequence has a duration of about five minutes).

2.3 C. STATS: Anonymized Data

The STATS dataset [10] consists of 7578 short game episodes (from 5 to 150 s), taken from 45 matches played in a top European league with total time of 2220 min. An individual episode starts when a certain team gets possession of the ball and ends when the team loses control of the ball.

Since each episode is anonymous, there is no way to extract both attacking and defending patterns of the same team or to obtain episodes where a specific team partic-ipates. Thus, we can treat this dataset as a "collective image" of a highly skilled human team. Like the DataStadium set, STATS does not include event markup, so passes, move-ments and shots on goal have to be reconstructed. One distinctive feature of this dataset is a low number of shots on goal. In most cases an episode ends when the ball is lost due to *any* reason including a shot on goal, so our options for analyzing episode outcomes are limited.

3 Team Behavior Fingerprinting

Before discussing possible approaches to fingerprint team behavior patterns, we have to make two preliminary notes. First, every team in subsequent experiments is treated as having its defensive zone on the right-hand side of the field. When processing left hand-side team data, we mirror all the coordinates to allow direct comparison between teams. Second, full-length matches from the DataStadium set were divided into halves to make possible to compare team behavior in the first and the second half of the game. We tried to split game recordings into smaller blocks (quarters), but it did not yield any significant changes in the results.

3.1 Ball Possession-Based Fingerprinting

The idea of a ball possession metric is to gather information about locations, where the player (belonging to the team of interest) controlling the ball spends time. We divide the soccer field divided into 64 cells (see Fig. 1).

On every frame of game recording, we increase the counter associated with a cell currently occupied by the player possessing the ball. We do not increase counters during passes (when the ball is not possessed by any particular player) and defensive actions (when the ball is possessed by the opposing team). At the end of this process we convert frequencies into percentages that can be visually represented as a heatmap (see Fig. 2). Such a heatmap can presumably reflect team-specific behavior patterns.

3.2 Pass-Based Fingerprinting

The idea of a pass/receive metric is similar to ball possession. Whenever a player per-forms a pass, we increase a counter for the cell occupied by the player. Similarly, we increase a counter for the cell where a pass is received (see Fig. 3). The logic behind this metric is a presumption that players have more freedom in choosing the targets of their passes than in their movements. Thus, passes may reflect players' tactical preferences more accurately.

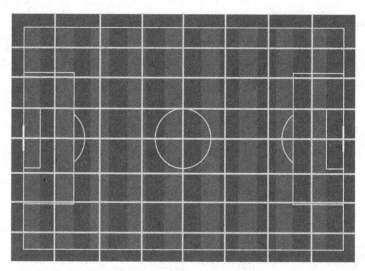

Fig. 1. Cells for ball possession analysis

3.47	4.41	5.90	6.64	1.92	0.62	1.02	0.71
0.43	0.07	1.69	0.43	0.87	0.87	0.13	0.12
0.51	0.95	1.52	0.83	0.42	0.48	2.20	0.31
0.05	0.66	1.40	1.98	1.79	0.81	0.62	4.62
0.04	0.78	1.37	0.85	0.58	1.59	0.67	2.16
0.12	0.00	0.75	0.12	1.12	0.94	1.52	2.41
0.07	0.42	0.98	0.47	2.31	0.83	0.00	0.00
5.30	4.84	6.96	5.27	3.98	3.01	2.07	0.03

Fig. 2. Sample heatmap of ball possession

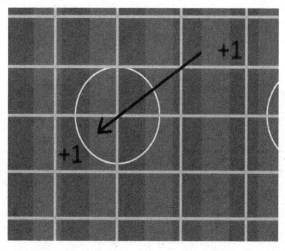

Fig. 3. Pass/receive frequency counting. Cells marked with +1 will get their counters increased

3.3 Heatmap Comparison

To compare heatmaps, we use a conventional cosine similarity measure, based on a dot product formula (Tables 1 and 2):

Table 1. Similarity of ball possession heatmaps (%)

	V11	V12	N11	N12	Y11	Y12	N21	N22	S11	S12	S21	S22	Y21	Y22	V21	V22	ST	SF
V12	78																	
N11	77	79																
N12	65	79	73															
Y11	81	80	80	79														
Y12	70	77	69	77	83													
N21	68	77	74	78	83	75												
N22	68	81	61	73	74	82	75											
S11	71	79	82	74	85	79	79	71										
S12	69	78	61	74	78	84	71	85	74									
S21	72	83	75	85	88	80	82	81	84	76								
S22	71	80	71	70	82	79	72	85	77	76	86							
Y21	67	71	84	75	82	71	73	58	82	63	76	72						
Y22	74	82	75	80	83	76	79	83	75	81	82	79	83					
V21	68	73	73	79	83	79	77	71	78	70	89	77	75	74				
V22	70	78	74	70	86	84	74	67	85	75	80	77	79	75	81			
ST	80	89	86	86	93	88	86	82	90	82	92	87	88	89	89	90		
SF	52	38	64	47	43	38	39	28	40	34	29	30	55	48	33	29	46	
WK	54	38	55	43	38	38	34	29	35	37	24	29	50	47	28	29	43	86

Table 2. Similarity of pass/receive heatmaps (%)

	V11	V12	N11	N12	Y11	Y12	N21	N22	S11	S12	S21	S22	Y21	Y22	V21	V22	ST	SF
V12	78																	
N11	74	86																
N12	77	76	72															
Y11	86	81	77	82														
Y12	80	77	68	79	81													
N21	74	80	79	84	83	74												
N22	75	74	65	72	73	73	74											
S11	80	79	79	81	80	87	84	76										
S12	79	85	75	79	79	81	78	83	86									
S21	83	86	82	84	87	83	88	73	89	83								
S22	85	82	74	79	88	85	72	78	79	82								
Y21	76	86	82	77	73	80	81	70	85	82	90	73						
Y22	75	87	75	77	78	72	79	83	75	83	83	79	80					
V21	70	68	65	82	79	70	82	66	77	72	77	70	69	67				
V22	80	81	75	77	82	86	77	73	86	86	86	78	82	75	64			
ST	88	92	83	87	90	90	87	82	92	90	94	91	90	88	81	89		
SF	57	50	47	50	43	39	46	49	45	46	39	43	50	43	46	40	54	
WK	61	54	53	59	48	49	52	61	51	55	46	52	55	52	51	47	62	93

4   Discussion

Ball possession has been chosen as the most straightforward approach and was supposed to provide a common-sense basic representation of a team style. However, the results show that most human teams follow similar behavior patterns, yielding similarity values in the range 64–90%. It is also seemingly hard to derive "team-specific" patterns from heatmap visualizations (see Fig. 4). Comparison of human teams' heatmaps obtained with the pass/receive metric yields more diverse values lying in a range 58–89%.

0.1	0.0	3.8	3.3	8.4	3.0	0.9	0.0
0.6	0.0	1.0	2.2	1.0	2.7	1.3	0.5
0.1	1.0	0.1	0.8	0.5	2.3	1.3	1.7
0.2	0.8	1.8	6.8	0.7	1.3	0.0	4.2
0.0	1.0	3.3	2.7	4.0	0.6	1.0	2.8
0.0	0.3	1.2	1.5	0.6	0.6	0.1	0.2
0.0	1.0	1.8	0.8	1.5	0.1	0.1	0.4
2.7	3.1	2.4	7.1	4.6	0.2	1.3	0.9
1.2	3.0	4.5	4.1	6.6	3.2	0.7	0.6
0.0	0.4	0.9	1.9	3.1	1.6	3.0	0.0
0.0	0.7	0.8	1.3	1.1	3.3	1.4	0.6
0.8	0.1	0.8	1.2	1.6	0.6	0.5	2.5
0.4	0.0	1.5	1.0	2.6	2.9	0.3	3.4
0.8	1.3	0.6	1.9	2.6	4.1	0.8	0.9
0.4	0.9	0.4	0.1	3.6	2.7	1.8	0.3
0.4	1.5	1.0	3.1	4.1	2.1	0.0	0.6

Fig. 4. Heatmaps of V11 (up) and N11 (down) (sim: 74%)

Judging from the tables, there are no clear patterns that can be used to distinguish one human team from another or identify the same team in a different half or a match. We can suggest some possible explanations for this result.

The most straightforward theory would be to presume that there are *indeed* no clearly identifiable styles among professional teams in modern soccer. While this idea may sound implausible, some specialists voice similar opinions. For example, a well-known Italian manager and a former player Roberto Mancini argued in 2007 that future advances in football would come from physical preparation of players rather than tactics [14]. The underlying reasoning is that the maturity of football as a game and widespread adoption of fresh ideas drive soccer to a certain global uniform style, where country-specific or team-specific innovations are unlikely *(ibid)*.

A possibly more likely explanation would be to treat the current approach as not nuanced enough to capture subtle differences between the teams rather than patterns typical for most soccer team. In this case we can only hope for more accurate results in follow-up studies.

It is also difficult to obtain reliable results due to relatively small data collections that we have for human teams. Unfortunately, tracking data is not easy to obtain, and even same-team datasets are uneven due to seasonal changes in team lineup, injuries, and diversity of opponents.

At the same time, teams of bots form an isolated subset, clearly distinguishable from human teams using both proposed methods. Similarity values between bot teams and human teams are low, and high between the different bot teams. As noted before, "bot teams" actually consist of the same rule-based players, having player with the ball as the only exception. Thus, we can only confirm that the particular AI system does not possess human-like behavior traits according to our calculations.

As a result, the proposed research questions can be answered as follows.

RQ1. We have limited success with a heatmap/cosine similarity-based evaluation algorithm. It provides reasonable results but fails to distinguish human teams in our dataset. While this method is simple, it could produce reliable player identification in boxing and tennis games [15, 16].

RQ2, RQ3. The answers are "no" under presumption that heatmap comparison is a reliable method to analyze team play style. We see that the same teams behave differently in different matches and/or match halves, while distinct teams often show very similar behavioral patterns. However, a more sophisticated method of evaluation might be required.

RQ4. Yes, at least for the AI system built into Google Research Football. It is hard to say how representative are the two bot teams present in our datasets. Since we mostly rely on passes and movements of the player with the ball, the similarity between SaltyFish and WeKick could have been expected to be lower. On the other hand, both these bots optimize performance and rely on other players' cooperation. Their behavior is described as "simple and reasonable" by the Google Research team [17], so we can suggest that a "typical" AI system should exhibit similar traits. Thus, human-likeness is not a natural property of a team of bots, and even our simple method can easily distinguish virtual teams from real teams.

5 Conclusion

This work is aimed to establish an experimental framework for believability and play style evaluation in the game of soccer. We have integrated several player tracking data sources and obtained a diverse collection consisting of anonymized game fragments, game recordings of known human teams, and game recordings of AI-based teams. We have also reconstructed game events, such as passes, movements, and shots on goal in the datasets where this information was missing.

This setup allows us to evaluate different comparison procedures. We aim to identify team behavior traits, reflecting team individuality, certain tactical and strategic patterns, distinguishing this team from other teams and persistent across matches. Our earlier experiments with other game genres such as tennis and fighting revealed such stable player-specific patterns. However, these studies were focused on computer games rather than real-world events, and were limited to player-vs-player genres where the concept of "team behavior" does not exist.

The identification of such team patterns has application in sports analytics, but our immediate goal is team game AI, where the task of designing a diverse set of virtual opponents, using different attacking strategies, and possessing different skills can be a major challenge. In 2013 Sicart observed that a computer game FIFA'12 is already highly realistic in terms of physics and animation but falls short in the AI department. He considers FIFA's AI system too deterministic, scripted and predictable [18].

Our present study relies on two simple heatmap-based methods of assessing behavior similarity. The first method builds a heatmap of ball possession, and the second method creates a similar heatmap of ball pass/receive events. Heatmaps are compared using a cosine similarity metrics.

This general approach has known limitations. For example, it treats all heatmap elements as independent and does not take into account higher-level factors such as team formation. Still, it proved to be a reliable behavior fingerprinting strategy in simpler games.

While in the present work we could not reveal clearly identifiable differences between individual teams or noticeable similarities in behavior of the same team in different matches, we succeeded in separating real teams from virtual teams. This result is, however, limited to Google Research Football's built-in AI system with individual players controlled by WeKick and SaltyFish AI. We tend to believe that other conventional soccer AI systems (such as the ones used in commercial games) will provide similar results, but this question needs further investigation.

The existence of clearly identifiable team-specific behavior patterns also remains an open question. The conventional understanding of soccer implies the existence of distinctive "national styles", which seem to be much more pronounced in the past [19]. However, the convergence of tactics and strategy in modern soccer is also recognized [14]. Thus, it is not clear whether our current results are caused primarily by the limitations of the methods we use or by the convergence of play styles exhibited by different teams.

References

1. Kitano, H., Asada, M., Kuniyoshi, Y., Noda, I., Osawai, E., Matsubara, H.: RoboCup: a challenge problem for AI and robotics. In: Kitano, H. (ed.) RoboCup 1997. LNCS, vol. 1395, pp. 1–19. Springer, Heidelberg (1998). https://doi.org/10.1007/3-540-64473-3_46
2. Bates, J.: The role of emotion in believable characters. Commun. ACM **37**, 122–125 (1994)
3. Carling, C.: Analysis of physical activity profiles when running with the ball in a professional soccer team. J. Sports Sci. **28**(3), 319–326 (2010)
4. Link, D., Hoernig, M.: Individual ball possession in soccer. PLoS ONE **12**(7), e0179953 (2017). https://doi.org/10.1371/journal.pone.0179953
5. Le, H.M., Carr, P., Yue, Y., Lucey, P.: Data-driven ghosting using deep imitation learning. In: Proceeding of the 11th MIT Sloan Sports Analytics Conference (2017)
6. Khaustov, V., Bogdan, G.M., Mozgovoy, M.: Pass in human style: learning soccer game patterns from spatiotemporal data. In: IEEE Conference on Games (2019)
7. Khaustov, V., Mozgovoy, M.: Learning believable player movement patterns from human data in a soccer game. In: 2020 22nd International Conference on Advanced Communication Technology (ICACT), pp. 91–93 (2020)

8. Kempe, M., Vogelbein, M., Memmert, D., Nopp, S.: Possession vs. direct play: evaluating tactical behavior in elite soccer. Int. J. Sports Sci. 4(6A), 35–41 (2014)
9. Hewitt, A., Greenham, G., Norton, K.: Game style in soccer: what is it and can we quantify it? Int. J. Perform. Anal. Sport 16(1), 355–372 (2016). https://doi.org/10.1080/24748668.2016.11868892
10. STATS.com, Soccer Dataset. https://www.stats.com/artificial-intelligence/
11. DataStadium Inc., Offical website. https://www.datastadium.co.jp/
12. Khaustov, V., Mozgovoy, M.: Recognizing events in spatiotemporal soccer data. Appl. Sci. 10(22), 8046 (2020)
13. Google Research, Google Research Football with Manchester City F.C. Competition. https://www.kaggle.com/c/google-football
14. Wilson, J.: Inverting the Pyramid: The History of Soccer Tactics. Nation Books, New York (2013)
15. Mozgovoy, M., Purgina, M., Umarov, I.: Believable self-learning AI for world of tennis. In: 2016 IEEE Conference on Computational Intelligence and Games, pp. 1–7 (2016)
16. Mozgovoy, M., Umarov, I.: Building a believable agent for a 3D boxing simulation game. In: Second International Conference on Computer Research and Development, pp. 46–50 (2010)
17. Kurach, K., et al.: Google research football: a novel reinforcement learning environment. In: AAAI, vol. 34, no. 04, pp. 4501–4510 (2020). https://doi.org/10.1609/aaai.v34i04.5878
18. Sicart, M.: A tale of two games: football and FIFA 12. In: Sports Videogames, pp. 40–57. Routledge (2013)
19. Jordan, A.: Soccer Politics. National Playing Styles. https://sites.duke.edu/wcwp/tournament-guides/olympic-football-2016-guide/team-playing-styles-in-soccer/

Cosmic Density Field Reconstruction with a Sparsity Prior Using Images of Distant Galaxies

Naoki Yoshida$^{(\boxtimes)}$ and Xiangchong Li

Kavli IPMU (WPI), The University of Tokyo, Kashiwa, Chiba 277-8552, Japan
naoki.yoshida@ipmu.jp

Abstract. Future astronomical observations are expected to deliver multi- peta byte images of stars, galaxies, and supernovae. Processing the sheer volume of imaging and spectroscopic data is technically challenging, and producing scientific outputs from the big data will remain a key task in the next decade. We develop novel methods based on modern machine learning and deep learning to analyze data from Subaru Hyper Suprime-Cam. In this contribution, we focus on reconstruction of cosmic density field. We use the observation of gravitational lensing effect that causes slight deformation of shapes of galaxies. The collective effect can be used to reconstruct the large-scale density distribution. Our novel technique assuming a sparsity prior allows to reconstruct the density field in full three dimensions. Statistical analysis of cosmic structure enables accurate determination of a few fundamental quantities called cosmological parameters that describe the contents and the evolution of the Universe.

Keywords: Sparse model · Image analysis · Bayesian inference

1 Introduction

Recent development of large ground-based and space-borne telescopes enabled astronomers to survey a large area of the sky with unprecedented sensitivity. Such "sky surveys" typically generate peta-byte volume data, that contain images of millions of stars, galaxies, and supernovae. Prompt analysis of the big imaging data is almost a must to produce high impact science outputs, and statistical analysis of wide-area images ultimately enable us to understand the evolution and constitute of the Universe.

In 2014, we assembled a team of researchers in astronomy, statistics, and computer science to analyze the high quality imaging data collected by Subaru Hyper Suprime-Cam (HSC). HSC can take a snapshot of a patch of the sky of 1.5 deg^2 in 15 min, and its operation over 300 nights generates about 1 peta byte of raw data. There are a variety of science objectives including detection of distant supernovae and blackholes. Machine-learning methods have already been applied to the first-year and the third year data released by HSC collaboration.

© Springer Nature Switzerland AG 2022
S. Sachdeva et al. (Eds.): BDA 2021, LNCS 13167, pp. 93–99, 2022.
https://doi.org/10.1007/978-3-030-96600-3_8

Our team contributed to automatic supernovae detection and classification, for example [6,9,10].

Making a large cosmic "map" is another important goal of the HSC survey. Typically, images of millions of galaxies are used to discern the intervening mass concentration via a physical effect called gravitational lensing. Recently we proposed a novel technique of density field reconstruction based on sparse modeling [5]. Conventional methods yield noisy estimates for a large-area density distribution [2], but our method achieves high-contrast reconstruction by imposing a sparsity prior to an effectively linear problem.

2   Scientific Background

In the last decade, there has been a steep increase of observational data in astronomy. Sky survey is a commonly used mode of observation that runs a telescope to scan over a wide area of the sky, instead of pointing to specific objects. This should be compared with our usual use of telescopes and binoculars; an amateur astronomer would like to point a telescope to the moon, Jupiter, or to a newly discovered comet. A typical survey simply accumulates data (images) of a large contiguous area and records the shapes and brightness of the objects such as stars, planets, and nebulae. Recent large survey projects yield images and often spectroscopic data of millions to billions of galaxies, and thus allow us to obtain a census of the Universe in a statistically complete manner. The well-known success of the Sloan Digital Sky Survey can be partly attributed to the sheer volume of data and by the quality and quantity of the scientific outputs [1]. Sky surveys turned cosmology to literally precision science.

Recent observations also revealed the existence of mysterious substances called dark energy and dark matter, which altogether constitute about 95% of the present-day Universe. Understanding the nature of the dark sector is now placed as a major goal of astronomy and cosmology, and here we find the need for even bigger sky surveys. There a variety of approaches toward this goal, and the following two are thought to be promising; observing distant supernovae and probing the large-scale matter distribution with gravitational lensing.

A particular class of supernovae called Type Ia serve as a standard candle in extragalactic astronomy. The brightness variation of a Type Ia SN can be used to measure the distance to it or to the host galaxy. To this end, rapid detection and classification are the key task and our recent studies yield impressive detection rate and classification performance [6,9,10].

The large-scale matter distribution over a length scale of one hundred million light-years is a powerful probe of the matter content of the universe. In this contribution, we aim at reconstructing the matter distribution in a large region of our universe, which is then used to infer the fundamental quantities called cosmological parameters by statistical analyses. This involves measurement of the shapes of numerous galaxies and solving a linear problem to determine the solution - the underlying density distribution - as described in the next section.

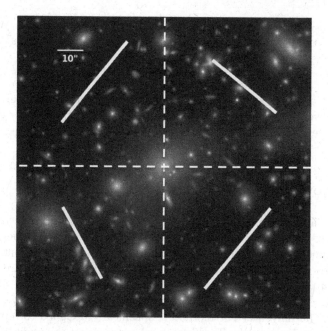

Fig. 1. Gravitational lensing effect. A massive galaxy at the center and its associated mass concentration deforms the space around it, so that the images of background galaxies appear slightly distorted. The collective effect is measured as a two-dimensional vector as depicted by the ticks in each portion. The observationally determined vector field can be used to reconstruct the mass distribution around the central massive galaxy.

3   Reconstruction of Matter Density Field

Gravitational lensing is a physical effect derived from Einstein's theory of general relativity. Mass concentration deforms the space-time around it, and light from distant galaxies is effectively bent near the gravitational source. The effect typically appears as a slight change of the overall shape of a background galaxy. Although the effect for individual galaxies is very small, the collective effect can be discerned by measuring the shapes of tens of galaxies within a small patch of the sky area. At each patch, there is a coherent direction of deformation and strength, that can be expressed as a vector. We call the vector as the lensing shear field (Fig. 1).

The lensing shear field γ is related to the foreground mass density field ρ via the lensing equation

$$\gamma = \mathbf{T} \cdot \delta + \epsilon \tag{1}$$

where

$$\delta = \frac{\rho}{\bar{\rho}} - 1 \tag{2}$$

with $\bar{\rho}$ denoting the mean density and ϵ the observational noise. We further assume that the density field is expressed as a sum of basis "atoms" as

$$\delta = \mathbf{\Phi} \cdot x \tag{3}$$

where the coefficient vector x is effectively the solution we seek for by solving the above reverse problem of δ to x. Note that, when appropriately discretized, Eq. (1) is essentially described as a linear problem with $\mathbf{T}$ expressing the matrix operator.

4   Previous Studies

Several studies attempted different dictionaries for the map reconstruction. Fourier space reconstruction represents the density field as a superposition of sinusoidal functions [8]. Unfortunately, a typical 3D density field contains a variety of modes with different wavenumbers and thus the solutions are not sparse in Fourier space. Starlet dictionary has been applied to many other applications [3], but it does not have a special advantage to model the clumpy density distribution in the universe. Our method is specifically designed to reproduce the realistic cosmic density field. To this end, we assume that the underlying density field is represented by a sum of the so-called NFW halos [7]. A NFW halo has a density profile of

$$\rho(r) = \frac{\rho_s}{\frac{r}{r_s}\left(1 + \frac{r}{r_s}\right)^2} \tag{4}$$

as a function of distance from the center r. It is known that nonlinear, gravitationally bound objects such as galaxies and galaxy clusters have density distribution that is accurately described by the NFW profile.

5   Sparsity Prior

A popular method to solve the linear problem of Eq. (1) adopts LASSO [11] by using

$$\hat{x}^{\text{LASSO}} = \text{argmin}_x \left\{ \frac{1}{2} \Sigma \|(\gamma - \mathbf{A} \cdot x)\|_2^2 + \lambda \|x\|_1^1 \right\}, \tag{5}$$

where Σ denotes the inverse of the covariance matrix of the shear measurements and $\mathbf{A} = \mathbf{T}\mathbf{\Phi}$.

The adaptive LASSO, on which we method is based, works in an iterative manner. First the standard LASSO is applied to estimate the coefficient vector. Then the result is used to calculate the weight vector as

$$\hat{w} = \frac{1}{\|\hat{x}^{\text{LASSO}}\|^\tau}. \tag{6}$$

The hyper-parameter τ sets the sensitivity to the weight, which we set to 2. Then the adaptive LASSO estimator is given by

$$\hat{x} = \text{argmin}_x \left\{ \frac{1}{2} \Sigma \|(\gamma - \mathbf{A} \cdot x)\|_2^2 + \lambda_a \|\hat{w} \cdot x\|_1^1 \right\}, \tag{7}$$

with the regularization parameter λ_a that sets the overall fluctuation level and smoothness of the reconstructed density field.

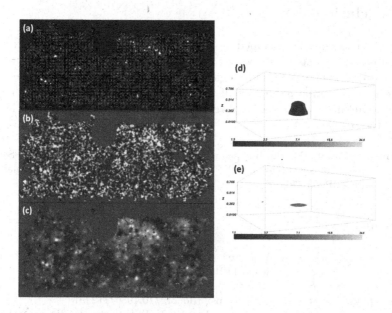

Fig. 2. Reconstruction of large-scale density field. (Left) For the same area of the sky of 30 deg^2, Panel (a) shows the cosmic shear vector field by white ticks, Panel (b) shows the reconstructed density field, and Panel (c) shows reconstruction with a sparsity prior. The color scale shows the density from white (highest density) to dark (low density). (Right) Three-dimensional reconstruction of the density field around a massive galaxy cluster. We compare the result of LASSO (d) and of our adaptive LASSO (e).

6   Application of Subaru HSC Data

Figure 2 shows the distribution of mass in a large area of the sky. We use data from HSC 16A. For the measured lensing shear field (Panel a), a conventional reconstruction method yields the density estimate with substantial noise (Panel b), but our sparsity-based reconstruction gives much smoother distribution with high effective signal-to-noise ratio. The results are promising, but the distribution is obtained on a projected two dimensional plane.

The adaptive LASSO performs well with three-dimensional reconstruction where the distribution in the line-of-sight direction is also accurately estimated. Panel (e) in Fig. 2 shows the distance uncertainty is substantially reduced by using the adaptive LASSO. We note that conventionally the line-of-sight separation (distance) is determined by identifying the member galaxies associated with the target mass concentration. The solution of our method automatically encodes the distance information in a direct manner [4].

7 Conclusion

We have successfully developed a multi-dimensional density reconstruction method based on a sparse estimator. We will apply the method to an updated catalogue that contains 40 million galaxies [4]. Future galaxy surveys will map a larger area covering about a half sky, and will deliver images of billions of galaxies. Our method can be applied to such big data to generate the biggest map of the cosmos.

Acknowledgement. The author acknowledges financial support by Japan Science and Technology Agency (JST) CREST JPMHCR1414, and by JST AIP Acceleration Research Grant JP20317829.

References

1. Gunn, J.E.: Jack of all. Ann. Rev. Astron. Astrophys. **58**, 1–25 (2020). https://doi.org/10.1146/annurev-astro-112119-041947
2. Kaiser, N., Squires, G.: Mapping the dark matter with weak gravitational lensing. Astrophys. J. **404**, 441 (1993). https://doi.org/10.1086/172297
3. Leonard, A., Lanusse, F., Starck, J.L.: GLIMPSE: accurate 3D weak lensing reconstructions using sparsity. Mon. Notice Royal Astron. Soc. **440**(2), 1281–1294 (2014). https://doi.org/10.1093/mnras/stu273
4. Li, X., et al.: The three-year shear catalog of the Subaru Hyper Suprime-Cam SSP Survey. arXiv e-prints arXiv:2107.00136, June 2021
5. Li, X., Yoshida, N., Oguri, M., Ikeda, S., Luo, W.: Three-dimensional reconstruction of weak-lensing mass maps with a sparsity prior. I. Cluster detection. Astrophys. J. **916**(2), 67 (2021). https://doi.org/10.3847/1538-4357/ac0625
6. Morii, M., et al.: Machine-learning selection of optical transients in the Subaru/Hyper Suprime-Cam survey. Pub. Astron. Soc. Jpn. **68**(6), 104 (2016). https://doi.org/10.1093/pasj/psw096
7. Navarro, J.F., Frenk, C.S., White, S.D.M.: A universal density profile from hierarchical clustering. Astrophys. J. **490**(2), 493–508 (1997). https://doi.org/10.1086/304888
8. Simon, P., Taylor, A.N., Hartlap, J.: Unfolding the matter distribution using three-dimensional weak gravitational lensing. Mon. Notice Royal Astron. Soc. **399**(1), 48–68 (2009). https://doi.org/10.1111/j.1365-2966.2009.15246.x
9. Takahashi, I., et al.: Photometric classification of Hyper Suprime-Cam transients using machine learning. Pub. Astron. Soc. Jpn. **72**(5), 89 (2020). https://doi.org/10.1093/pasj/psaa082

10. Yasuda, N., et al.: The hyper Suprime-Cam SSP transient survey in COSMOS: overview. Pub. Astron. Soc. Jpn. **71**(4), 74 (2019). https://doi.org/10.1093/pasj/psz050

11. Zou, H.: The adaptive lasso and its oracle properties. J. Am. Stat. Assoc. **101**, 1418–1429 (2006). https://EconPapers.repec.org/RePEc:bes:jnlasa:v:101:y:2006:p:1418--1429

Big Data Analytics in Domain-Specific Systems

Blockchain-Enabled System for Interoperable Healthcare

Vijayant Pawar[✉], Anil Kumar Patel, and Shelly Sachdeva

Department of Computer Science and Engineering, National Institute of Technology Delhi,
New Delhi, India
{vijayantpawar,202211003,shellysachdeva}@nitdelhi.ac.in

Abstract. Lack of interoperability and standardization of electronic health records (EHRs) are the most significant issues for effective EHR sharing. We have proposed a blockchain-enabled interoperable healthcare system that counters two problems regarding EHR context: first, how patients can have a complete view of their lifelong data to get the precise treatment and second, how doctors can get the real-time data of their patient in a standardized form which is interoperable between practitioners. The proposed solution follows the standards of openEHR to ensure the standardization of electronic records. It also offers fine-grained access permissions to the system's stakeholders (i.e., patient, doctor, insurance company, etc.). It uses archetypes of openEHR as a reference for making the templates and then these templates are integrated to build a module in compliant standards. The proposed solution uses Proof of Authority consensus (PoA) to tackle performance issues. The proposed system ensures data interoperability and the privacy protection of sensitive medical data.

Keywords: Blockchain · Electronic health records (EHRs) · openEHR · Interoperability

1 Introduction

Healthcare is one of the biggest and fastest-growing sectors across the world. Different organizations and professionals like hospitals, research labs, physicians, nursing homes, diagnostic laboratories, pharmacies and medical device manufacturers are involved in this and generate massive amounts of data. So, digitalization and analysis of healthcare data is the prime focus nowadays. But the digitization of health records brings a lot of challenges like data collection and cleaning, storage, security, querying, reporting and visualizations, sharing or exchanging data between stakeholders. We have focused on the interoperability, standardization and ownership of the patients' health records using a blockchain-enabled interoperable healthcare system. Many international organizations are working on standardizing clinical data to support interoperability. We proposed a blockchain-enabled system to tackle the problems of interoperability, lack of standardization while giving full ownership of the sensitive health care data to its owners (i.e., patients). Some of the specific terms discussed in the paper are discussed below:

© Springer Nature Switzerland AG 2022
S. Sachdeva et al. (Eds.): BDA 2021, LNCS 13167, pp. 103–121, 2022.
https://doi.org/10.1007/978-3-030-96600-3_9

1.1 Electronic Health Record

An electronic health record (EHR) is an encyclopedic digital record of a patient's medical history [1]. It is an integrated report from various sources and contains each and every minute detail of medical history from birth until the end of life.

Table 1. Standards related to electronic health records

Standards	Description
HL7/CDA [3]	The HL7 (Health Level 7) Version 3 Clinical Document Architecture (CDA) is a document markup standard, and it works on the structure and semantics of "clinical documents" in order to promote interoperability and promotes the exchange of medical data between various healthcare vendors
FHIR [4]	FHIR (Fast Healthcare Interoperability Resources) is a standard that is determined for data formats and elements and provides an application interface for promoting and exchanging electronic health records. The standard is created and maintained by the Health Level Seven International (HL7) healthcare standards organization
DICOM [5]	DICOM (Digital Imaging and Communications in Medicine) is the international standard for medical images and related information. It defines the formats for medical images that can be exchanged with the data and quality necessary for clinical use
CEN's TC/251 [6]	CEN/TC 251 (European Committee for Standardization (CEN) Technical Committee 251) is a technical decision-making committee working on standardization in the field of Health Information and Communications Technology (ICT) in the European Union
ISO-ISO TC 215 [7]	The ISO/TC 215 is the International Organization for Standardization's (ISO) Technical Committee (TC) on health informatics. TC 215 works on the standardization of Health Information and Communications Technology (ICT) to allow for compatibility and interoperability between independent systems
Continuity of Care Record [8]	The Continuity of Care Record, or CCR, is a standard for the creation of electronic summaries of patient health. Its aim is to improve the quality of health care and to reduce medical errors by making current information readily available to physicians
openEHR [9]	openEHR is a technology for e-health consisting of open platform specifications, clinical models and software that together define a domain-driven information systems platform for healthcare and medical research

Healthcare data is quite complex because it contains data of varied nature. Data in healthcare can be structured, unstructured and semi-structured. Lack of interoperability and standardization in electronic health records have always been an obstacle in healthcare. One of the main reasons for this was the non-availability of proper standards also non-acceptance of existing standards.

Table 2. Comparison of openEHR standards with other standards

Standards & open specifications	Pros	Cons
HL7/CDA	• XML is capable of supporting unstructured texts • XML is open and platform - independent standard	• Dedicated to XML based markup standards
FHIR	• Easy implementation uses web-based application program interface (API) • Multiple data representation choices. (JSON, XML, RDF)	• Inconsistencies because of improper implementation of APIs
DICOM	• Efficient transmission and storage of medical images • Integration of medical images from various sources	• File format allows executable code which may contain malwares
CEN's TC/251	• Promotes interoperability and compatibility between independent systems • Aims to provide architectures, framework and concepts for consistent and coherent implementations	• Lack of collaboration with international bodies • Based on the informatics interoperability within Europe only
ISO-ISO TC 215	• Promotes standardization of health and communication technology • Dedicated working groups for various aspects of health records	• There can be disagreement in standards set by different committees
Continuity of Care Record	• XML based standard data interchange • Involvement of patient in data management	• Security is an issue without authorized data transmission • Inclined more towards text data
openEHR	• Availability of standard archetypes/templates • Allows querying based on archetypes (AQL) • Archetypes and templates can be modified as per need	• Focuses more on data persistence, data exchange is secondary focus • Expertise is required to handle archetypes and templates

Acceptance of any standard by an organization is also influenced by rules and regulations imposed by the government of their country. In Table 1, we have listed the set of globally accepted standards and specifications related to EHRs such as HL7/CDA [3], FHIR [4], DICOM [5], CEN's TC/251 [6], ISO/TC 215 [7], CCR [8], openEHR [9]. Standard like CEN/TC 251 is the standard set for European Union and is accepted and followed in European Union countries.

Digital Imaging and Communications in Medicine (DICOM) is specifically a standard related to medical imaging and information. Standards like Health Level 7 are used for Clinical Document Architecture (CDA) and FHIR (Fast Healthcare Interoperability Resources) is maintained by a non-profit organization dedicated to providing standards and solutions promoting global health interoperability. The ISO/TC 215 Technical Committee (TC) on health informatics is part of the International Organization for Standardization (ISO). TC 215 is responsible for standardizing health information and communication technology (ICT) to ensure system compatibility and interoperability.

Table 2 presents the pros and cons of the openEHR standard with other standards. openEHR provides the specification for open standards-based clinical models. It also offers repositories and tools like clinical knowledge manager and archetype designer. The most significant advantage of using openEHR specifications is the reusability of the clinical data repository. In the openEHR repository, there are more than 250 archetypes and templates that can be used to build digital patient records and applications.

1.2 The Standard for Semantic Interoperability

The three predominant standards for semantic interoperability are HL7/CDA, openEHR and ISO/TC 215. Based on the pros and cons presented in Table 2, the current research considers the openEHR standard for maintaining and sharing health record data. openEHR standard follows a dual model approach consisting of reference model (RM) and domain-level descriptions in the form of archetypes and templates.

Archetypes: An archetype [2] provides structure to the system. openEHR has predefined archetypes which can be exported directly in XML format. For example, take the archetypes of blood pressure from openEHR (https://ckm.openehr.org/ckm/). These are reusable models of a domain concept. It is modifiable and can be used as per the requirement. It provides a blueprint for capturing clinical data and is used for making templates.

Templates: A template agglomerates archetype or group of archetypes as per the requirement of the system. A template is usually a visual interface created for the ease of the user, which the end-user can directly use for data collection or for the generation of printed reports. Data in the database can be stored using templates and retrieved using queries. By integrating archetypes, we can create standard templates.

1.3 Blockchain

Blockchain is a distributed ledger that stores an immutable record of transactions to enable decentralized and secure transactions. The transactions are collected into a block, which is then linked to the chain. Initially, blockchain technology was used to facilitate the exchange of digital money packaged as transactions that are recorded on an immutable, tamper-proof ledger and exchanged through a peer-to-peer network. It's worth emphasizing that while these technologies aren't new, their application together creates blockchain, which is a new technology. We define the blockchain as an amalgamation of distributed ledger, consensus algorithms and cryptography.

$$\int (\text{Distributed ledger, Consensus algorithm, Cryptography}) \tag{1}$$

The parameters of Eq. (1) are described as follows:

Distributed Ledger: The distributed nature of blockchain technology makes the system fault-tolerant by eliminating the need for a central authority. In a blockchain network, each node holds the exact copy of the ledger. If any node behaves maliciously or falsely, copies of the blockchain can be retrieved from other system nodes. This improves the fault tolerance capability of the system.

Consensus Algorithm: The consensus algorithm is responsible for appending any block to the blockchain network. In a blockchain network, if the number of nodes is agreed on a single state of truth that is higher than a threshold value (i.e., 51% for PoW), then that particular transaction is allowed to append to the blockchain.

Cryptography: On the blockchain network, the cryptography technique generates a secure digital identity for each participant and validates transactions. This is accomplished by applying hashing and participants' public and private keys.

Blockchain is the emerging technology that provides transparency, immutability, privacy (pseudo-anonymity) and traceability to healthcare data. This paper presented a blockchain-enabled openEHR compliant standard solution to interoperate the healthcare ecosystem. To the best of the author's knowledge, no blockchain-enabled solution follows openEHR standard compliant for healthcare data.

1.4 Research Contributions

In this work, we have proposed a blockchain-enabled interoperable healthcare system that uses openEHR standards for healthcare data sharing. Compared with state-of-the-art results, the contribution of the article is listed as follows:

- It aims to build an architecture for a blockchain-enabled system for interoperable healthcare.
- It is a complaint to openEHR standard, a standard for semantic interoperability of electronic health records.
- It presents a case study of blockchain-based health insurance with implementation details.

The paper is organized as follows. Section 2 summarizes most related works. Section 3 discusses the architecture of a blockchain-enabled interoperable healthcare system. In Sect. 4, we cover the implementation details of template generation. In Sect. 5, a blockchain-based health insurance use case is discussed with implementation details. We conclude this paper with a future research discussion in Sect. 6.

2 Related Work

Healthcare data is regarded as highly sensitive and requires a safe and reliable means to protect it. Therefore, medical data should be collected, exchanged and handled securely [11, 12]. There are already numerous frameworks proposed to resolve these issues; for example, the schemes set out in [13–15] to meet the need for safe and efficient accessibility, manageability and other critical security criteria for medical data. To some extent, these solutions help offer different safety requirements under desired healthcare scenarios. However, with the latest developments in healthcare technology, these strategies are not enough. Various stakeholders have exploited the patient through unfair means and they access patient data without their consent [16, 17]. Therefore, researchers are keen to find various security solutions in the context of blockchain-based approaches to healthcare [18]. Numerous research studies have been carried out in [19, 20] regarding blockchain's potential use in healthcare. Electronic medical treatment processes for manual and remote access to patient data and safeguarding healthcare data privacy are the most foregoing application areas where blockchain technology can create value [21].

SimplyVital Health [22] is a startup company founded in 2016 aiming to leverage blockchain technology in healthcare. Their publicly available web materials indicate working on two products: "ConnectingCare" and "Health Nexus." ConnectingCare is a care coordination tool used by providers that leverage blockchain technology to create a secure audit trail. Health Nexus plans to be its healthcare-focused blockchain with an underlying cryptocurrency token called "HLTH." Medrec has proposed the study in [23], in which a decentralized approach of using blockchain technology to handle the EHR/EMR (Electronic Medical Record) is adopted. The authors have presented a possible case study of blockchain usage in healthcare that offers an EHR/EMR prototype. The various challenges and opportunities like Clinical Health Data Exchange, Interoperability, Claims Adjudication, Patient Billing Management and Drug Supply Chain Integrity that blockchain applications face in the healthcare industry are discussed [24]. The authors believe blockchain technology will improve patient care, treatment efficacy, security and reduce costs. They suggest that electronic medical record management will be more efficient, disintermediated and secured through blockchain technology.

Kuo et al. [25] provide a comprehensive overview of biomedical and healthcare applications that could be developed using blockchain technology. They view decentralized management of records, the immutability of audit trails, data provenance, data availability, security and privacy as benefits from implementing blockchain technology over traditional distributed database management systems. The authors also discuss potential challenges that blockchain applications may face in a healthcare environment relating to transparency, confidentiality, speed, scalability and resistance to malicious actors. Metrics can be used to evaluate the feasibility, intended capability and compliance of a blockchain-based application in the healthcare space discussed in [26]. Dufel

[27] discusses blockchain alongside other peer-to-peer technologies and their application to healthcare. The authors suggest that only a combination of blockchain, distributed hash tables and BitTorrent would effectively create a peer-to-peer health information exchange system.

3 Blockchain-Enabled System for Interoperable Healthcare

The proposed architecture of blockchain-enabled healthcare architecture is presented in Fig. 1, which consists of three different modules. The first module is the stakeholder's module, followed by the blockchain and storage module. The description of these modules are as follows:

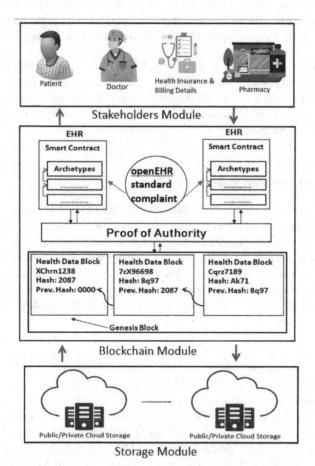

Fig. 1. Modular architecture of blockchain-enabled system for interoperable healthcare

Stakeholders' Module: This module contains various stakeholders of the system such as patients, doctors, pharmacy, specialists, insurance company etc. which interact with the system to avail various services of the blockchain-enabled healthcare system.

Blockchain Module: This module is responsible for storing the data into the storage module while following openEHR standards. It uses smart contracts to verify whether the data which is requested to be stored in the storage module is in openEHR standard or not. Smart contracts are simply programs stored on a blockchain that run when predetermined conditions are met. We use the proof of authority consensus mechanism to append metadata to the blockchain. Smart contracts are also used to control data accessibility. If any stakeholder of the system wants to access other stakeholders' data first, he will have to request access to the data owner and if the owner allows, then only access will be granted. The data owners can revoke access at any time. It is not feasible to store each and every bit of data on the blockchain; thus, only the metadata of the health records are stored on the blockchain and actual data is stored on the cloud storage.

Storage Module: The storage module is used for storing EHRs in the openEHR standard. Due to the huge data volume, each and every detail of every patient, such as lab reports, doctor prescriptions, health insurance policy details, pharmacy details etc., is not possible to store in the hospital database. Thus, a remote location, i.e., public/private cloud, is used to store the data.

3.1 Standard Complaint System

This section proposed a standard complaint blockchain-enabled system to achieve interoperability amongst different healthcare providers for sharing clinical data. Earlier FHIRChain [28], a blockchain-based architecture designed to meet ONC requirements by encapsulating the HL7 Fast Healthcare Interoperability Resources (FHIR) standard for shared clinical data, was proposed. The current study has proposed an openEHR standard complaint system to achieve data interoperability. Since openEHR uses archetypes for semantic interoperability. As shown in Fig. 1 the blockchain module checks whether the data which is received from the stakeholder's module is following the standard or not via smart contract. If consensus is reached over the data format, the related metadata is appended to the blockchain. The entities which make the data standard-compliant is discussed in the below subsection.

Clinical Knowledge Manager (CKM)
This subsection explains the importance of CKM and presents the snapshots for creating an openEHR compliant system. openEHR has a well-developed system called Clinical Knowledge Manager [10]. Clinical Knowledge Manager contains a lot of predesigned archetypes and templates. These archetypes can be exported in ADL or XML format. Before selecting any archetypes, it's necessary to get the complete details about them, like whether it has all the required parameters or not. It contains all the archetypes,

templates and other resources for clinical models. CKM is a system used to develop, upgrade, manage and release archetypes and templates.

Archetype and Template Generator

Clinical Knowledge Manager has more than two hundred plus archetypes for various purposes like observation, instruction, evaluation, test results etc. There is a possibility that the existing archetypes and templates are not following the user system requirements. OpenEHR Archetype Designer [29] solves this issue of the unavailability of needed archetypes and templates. Archetype Designer is a platform that allows the creation of archetypes and templates from scratch. Also, existing resources can be modified using Archetype Designer. Once templates (formed using the integration of various archetypes) have been designed, they can be exported in various languages and formats. The most popular web-based template is the.opt format which stands for operational-template.

Graphical User Interface Generation

End-users of electronic health records are patients, doctors, and other vendors. So, it becomes very important that the templates being created must be highly user-friendly. After exporting the templates from the archetype designer, the next step is to create user-friendly forms. These can be created either using the template's metadata or by directly using the tools like Medblocks UI Extension [30] or openEHR Cabolabs Toolkit [31].

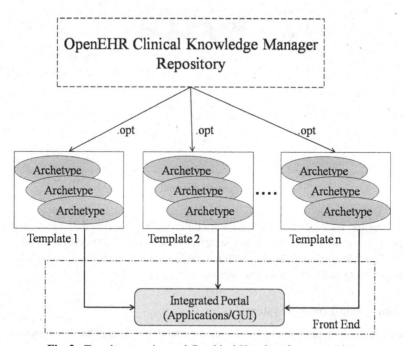

Fig. 2. Template creation and Graphical User Interface generation

Figure 2 shows the process of template creation and graphical user interface generation. The templates are exported in operation template format (.opt) from the clinical knowledge manager repository. There are basically two ways for designing and developing the templates. The first method is to create a template from scratch using a collection of archetypes. The second method is to check if a template is already available in the CKM repository. If it is available, it can be extracted in the required format and integrated to build the graphical user interface (GUI). If a template is not available as per the need, existing templates can also be modified and some archetypes can also be added to fulfill the requirements. openEHR provides one more representation of archetypes called "Archetype Mind Map". Figure 3 contains a mind map view of the laboratory test results archetype. It provides a clear overview of an archetype and its various attributes.

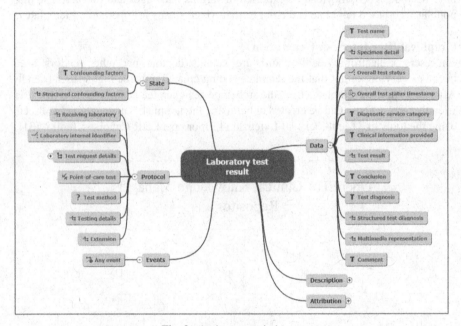

Fig. 3. Archetypes mind map

The components of the blockchain module, such as health data block, Proof of Authority consensus mechanism and smart contracts, are discussed in the preceding section. To achieve interoperability, manual creation of HTML forms and modules using archetypes metadata is discussed in Sect. 4.

3.2   Blockchain-Enabled System

We leverage the capabilities of blockchain to make healthcare data interoperable and patient-controlled. Blockchain is responsible for storing the data into the storage module while following openEHR standards. It uses smart contracts to verify whether the data requested to be stored in the storage module follows the openEHR standard. Various essential components of the blockchain module are described as follows:

Health Data Block: A block is a fundamental unit of blockchain that contains data. A block in the blockchain is linked to its predecessor block through the hash value of the predecessor block. A block contains data like previous hash, current hash, timestamp, metadata as a reference of the actual data stored in the cloud storage.

Proof of Authority (PoA): Proof of Authority is a consensus algorithm for achieving consensus in the distributed setting. It is a type of Proof of Stake protocol in which the node's identity is kept at risk instead of monetary wealth. Unlike Proof of Work, a blockchain network that uses Proof of Authority consensus does not require significant computational processing. In PoA, consensus algorithm validators are responsible for safeguarding the network. Validators are picked democratically while using PoA. The network's nodes are incentivized to validate the network's transactions. When the validator validates a fraud or harmful transaction, its reputation gets hampered for doing the fraudulent act.

Smart Contract: Like any other contract, a smart contract lays forth the parameters of a deal. However, unlike a typical contract, the rules of a smart contract are carried out as code on a blockchain. Developers may use smart contracts to create apps that utilize blockchain security, stability, and accessibility while also providing decentralized capabilities.

To facilitate the interaction between various stakeholders of the blockchain-enabled healthcare system. Ethereum and GPG Public/Private key pair is generated as shown in Fig. 8 of Sect. 4.2. The complete process of applying blockchain technology for health insurance is presented in Fig. 9 and the health insurance process is supported with the snapshots which are shown in Fig. 10 and Fig. 11.

4   Results

This section provides complete information on archetypes, templates and HTML form generation. Detailed implementation of every component is covered in this section. We also discuss the creation of GPG keys for the authorized EHRs access in the health insurance case study.

4.1   Standard Complaint System

This section presented a standard-compliant blockchain-enabled system to establish interoperability across healthcare providers for clinical data sharing. It consists of implementation details of template and archetype generation, HTML form generation using openEHR toolkit and creation of forms and modules using archetypes metadata.

Archetypes and Template Generation

This section provides implementation details of archetypes and template generation. Once an archetype is exported, it needs to be modified according to the user's needs and requirements. Some attributes need to be removed or appended to fulfill the user requirements. For all these operations, openEHR provides "Archetype Designer".

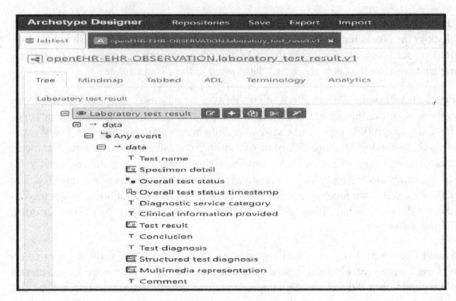

Fig. 4. openEHR archetype designer.

Archetype Designer supports the ADL format of archetypes. It also allows creating new archetypes, using tools provided in the Archetype Designer, new templates and archetypes to be created and modified. In Fig. 4 template for laboratory test results is being designed.

Once all the necessary changes have been made in the template and the final template is ready, the next step is to export it for further GUI creation. In Fig. 5, it can be seen the various language and export formats available depending on the region.

HTML form Generator Using openEHR Toolkit
Graphical User Interface (GUI) form can either be generated by using openEHR toolkit, UI Extensions such as "Medblocks UI Extension" or can be developed manually. As of now, there's not any direct method for converting the openEHR template (OPT) directly into HTML forms.

But Cabolabs openEHR Toolkit [31] provides "HTML Form Generator". The operational template (OPT) is provided as input in this tool, which generates the HTML form. Figure 6 shows the HTML form corresponding to the template designed previously and exported in .opt format.

Manual Creation of HTML Forms and Modules Using Archetypes Metadata
We can also create an HTML form from the metadata in the OPT format. This is a complex operation and needs a proper understanding of openEHR archetypes and OPT model. Figure 7 shows the result of the web form creation using metadata of OPT.

Similarly, other templates can be created as well and can be integrated for the development of other modules of the EHR system. Once all the modules are developed, the next step is linking it with Ethereum blockchain via smart contracts to develop a fully functional openEHR compliant system for clinical data sharing.

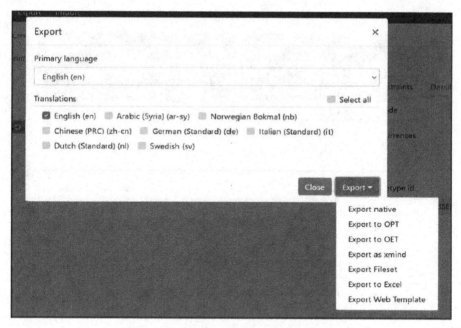

Fig. 5. Exporting templates in archetype designer

Fig. 6. Cabolabs toolkit HTML form generator

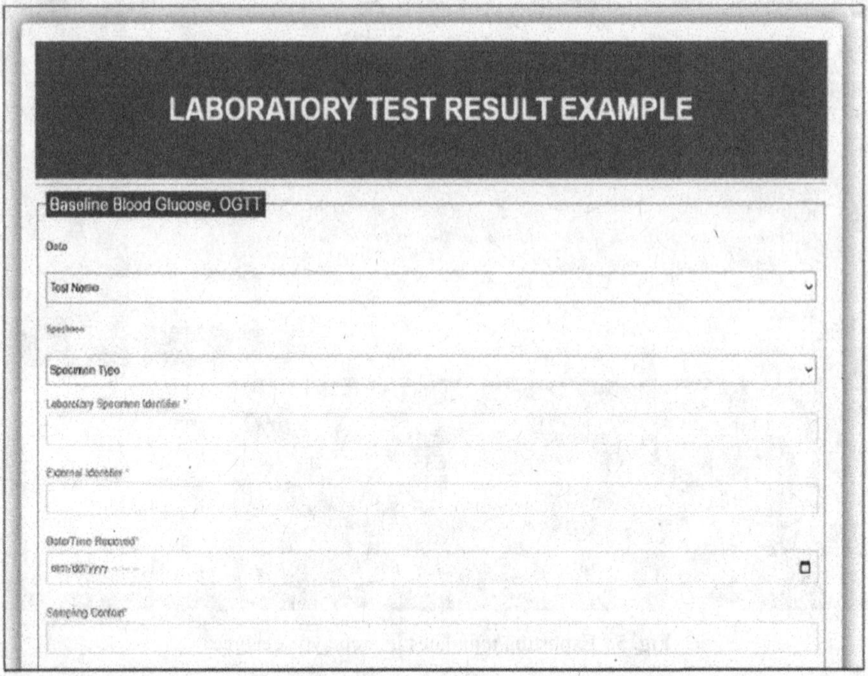

Fig. 7. HTML form using archetypes metadata

4.2 Blockchain-Enabled System

This section has discussed the blockchain module of the proposed blockchain-enabled healthcare system. We also present a blockchain use case in the health insurance sector which shows various transactions between insurance agencies, patients and doctors. Ethereum and GPG Public/Private key pair is generated to facilitate the interaction between various stakeholders, as shown in Fig. 8. The complete process of applying blockchain technology for health insurance is demonstrated in Fig. 9. The key generation for the health insurance process is shown below:

1. A Public/Private key pair (PubKU, PriKU) is generated. The private key is generated using SHA256. Then, the Elliptic Curve Digital Signature Algorithm (ECDSA) and secp256k1 operations are performed to generate the public key corresponding to the generated private key. The generated (PubKU, PriKU) key pair is used for authentication of stakeholders to interact with the blockchain module.
2. GPG key pair (GPGPubKU, GPGPriKU) is also created, which is used for encrypting and decrypting the data. The key pair is generated using the GnuPG standard along with a strong paraphrase to protect the keys. For the Ethereum key pair, the framework leverages the Web3.js package to build externally owned accounts for each system entity. Similarly, the GnuPG.js package is used for encryption and decryption of data that is sent to all the involved stakeholders of the system. Figure 8 provides a

code sample for producing GPG keys for a patient. Similarly, the key pair is also generated for other stakeholders of the system.

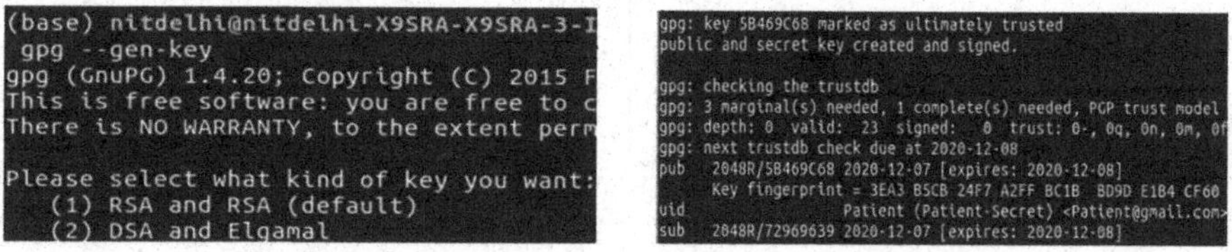

Fig. 8. Generation of GPG keys for the patient

5 Case Study: Blockchain-Based Health Insurance System

The healthcare cost in developing countries such as India is very high, stringent and time-consuming. The main concern in the traditional healthcare system is the slow speed of the medical reimbursement process. The flawless and speedy reimbursement process allows doctors to proceed with the treatment without waiting for the insurers or payers to respond. The health insurance company can store their policies through smart contracts onto the blockchain. If a person wants to buy a particular health policy, his information is mapped to a policy via a smart contract and stored onto the blockchain. In the case of medical urgency, the patient submits an application to the blockchain for prior authorization from the insurance company. The smart contract for the patient's medical policy specifies whether the treatment is covered under the policy or not. Then, authorization data will then be sent to the healthcare provider immediately. The patient and any stakeholder like pharmacy, laboratory, or consultant doctor could check in real-time for insurance authorization. This will help in providing hassle-free and real-time treatment to the patient.

Figure 9 shows the working of a blockchain-based health insurance system. Blockchain-based health insurance system consists of six steps: policy creation, stakeholders' registration, procedure requested, checking whether the procedure is registered or not, procedure executed and fund debited, and procedure and fund verification.

Policy Creation: The insurance agency is responsible for policy creation. The insurance administrator specifies the policy type, benefits associated, excluded diseases and decide the policy renewal date. The admin insurer then designs and deploys a policy smart contract on the blockchain.

Stakeholder Registration: When a health policy is created, the insurance supervisor must have a list of eligible patients and doctors.

The insurance supervisor can authenticate and authorize eligible stakeholders using the (PubKU, PriKU) keys. If a person is insured, they are given a secure digital key. After that, for each eligible investor, a wallet is created. Each shareholder has a wallet, which can access the insurer's services.

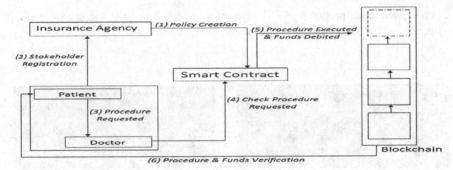

Fig. 9. A use case of applying blockchain technology for health insurance

Procedure Requested: The patient requests the doctor to execute the needed procedure.

Procedure Validation: When a doctor receives a request from the patient for a particular treatment. The doctor checks whether the procedure is covered for the patient or not via a smart contract. The smart contract only processes legitimate requests, indicating that the policy covers the required operation.

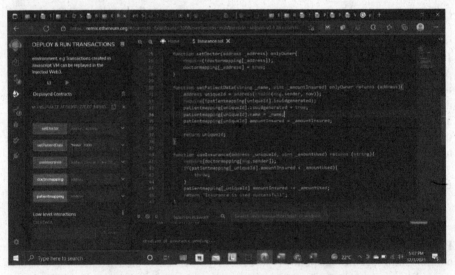

Fig. 10. Code snippet for patient insurance module

If the procedure follows the policy guidelines, the smart contract sends the procedure request transaction to all blockchain network nodes. Then the procedure request transaction is permanently appended to the blockchain, and the transaction ID is provided to the patient and doctor. Consensus for the specific transaction is obtained when the majority of blockchain nodes agree on the joint decision.

Procedure Executed and Fund Debited: According to the procedure requested transactions stored on the blockchain, the policy smart contract automatically debits the fund allotted for that operation.

Procedure and Fund Verification: Once the procedure is executed, confirmation and fund debited notifications are sent to the doctor and patient.

The code snippet for the patient insurance module and transaction logs are shown with the help of Fig. 10 and Fig. 11.

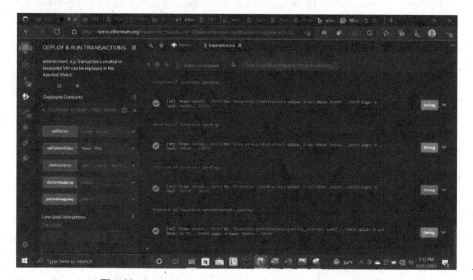

Fig. 11. Logs of the transactions created in the blockchain

6   Conclusions

This article examines how we can improve the existing healthcare system by bringing interoperability to the data and providing more authority to the patient over the data related to its medication lifecycle. The proposed system uses openEHR standards for achieving data interoperability. The blockchain-based system allows users to access sensitive medical data in a user-centric manner. Implementing such a system in the actual world would provide several benefits to all concerned stakeholders. Patients should manage their medical analyses easily, and hospitals should give better medical care; health insurance companies and research organizations have simple access to a significant health database. The proposed framework's effectiveness is demonstrated by deploying an insurance use case on the permissionless blockchain (i.e. ethereum). Furthermore, we believe that patients should use micro-transactions to allow access to limited views of their data. We'll look into ways to enable safe health data interchange among many IoT devices authorized via a blockchain network in future work.

References

1. Kettley, P., Reilly, P.: eHR: An Introduction. IES Report. Grantham Book Services, Ltd., July 2003. http://www.employment-studies.co.uk
2. Archetypes. https://specifications.openehr.org/releases/1.0.1/html/architecture/overview/Output/archetyping.html. Accessed 10 Nov 2021
3. HL7. CDA. Clinical document architecture. Release 2. http://www.hl7.org/v3ballot/html/infrastructure/cda/cda.htm
4. Bender, D., Sartipi, K.: HL7 FHIR: an Agile and RESTful approach to healthcare information exchange. In: Proceedings of the 26th IEEE International Symposium on Computer-Based Medical Systems, 20 June 2013, pp. 326–331. IEEE (2013)
5. DICOM: Digital imaging and communications in medicine. https://www.dicomstandard.org
6. CEN TC/251: European standardization of health informatics. ENV 13606 Electronic Health Record Communication. http://www.centc251.org/
7. ISO/TC 215 Technical report: electronic health record definition, scope, and context. 2nd draft, August 2003
8. Ferranti, J.M., Musser, R.C., Kawamoto, K., Hammond, W.E.: The clinical document architecture and the continuity of care record: a critical analysis. J. Am. Med. Inform. Assoc. **13**(3), 245–252 (2006). https://doi.org/10.1197/jamia.M1963
9. https://www.openehr.org. Accessed 10 Nov 2021
10. Clinical Knowledge Manager. https://ckm.openehr.org/ckm/. Accessed 10 Nov 2021
11. Puppala, M., He, T., Yu, X., Chen, S., Ogunti, R., Wong, S.T.C.: Data security and privacy management in healthcare applications and clinical data warehouse environment. In: 2016 IEEE-EMBS International Conference on Biomedical and Health Informatics (BHI), pp. 5–8, February 2016
12. Abouelmehdi, K., Beni-Hssane, A., Khaloufi, H., Saadi, M.: Big data security and privacy in healthcare: a review. Procedia Comput. Sci. **113**, 73–80 (2017). The 8th International Conference on Emerging Ubiquitous Systems and Pervasive Networks (EUSPN 2017)/The 7th International Conference on Current and Future Trends of Information and Communication Technologies in Healthcare (ICTH-2017)/Affiliated Workshops
13. Kahani, N., Elgazzar, K., Cordy, J.R.: Authentication and access control in e-health systems in the cloud. In: 2016 IEEE 2nd International Conference on Big Data Security on Cloud (BigDataSecurity), IEEE International Conference on High Performance and Smart Computing (HPSC), and IEEE International Conference on Intelligent Data and Security (IDS), pp. 13–23, April 2016
14. Azeta, A.A., Iboroma, D.O.A., Azeta, V.I., Igbekele, E.O., Fatinikun, D.O., Ekpunobi, E.: Implementing a medical record system with biometrics authentication in E-health. In: 2017 IEEE AFRICON, pp. 979–983, September 2017
15. Kumar, T., Braeken, A., Liyanage, M., Ylianttila, M.: Identity privacy preserving biometric based authentication scheme for naked healthcare environment. In: 2017 IEEE International Conference on Communications (ICC), pp. 1–7, May 2017
16. Yüksel, B., Küpçü, A., Özkasap, Ö.: Research issues for privacy and security of electronic health services. Future Gener. Comput. Syst. **68**, 1–13 (2017). http://www.sciencedirect.com/science/article/pii/S0167739X16302667
17. Jabeen, F., Hamid, Z., Akhunzada, A., Abdul, W., Ghouzali, S.: Trust and reputation management in healthcare systems: taxonomy, requirements and open issues. IEEE Access **6**, 17246–17263 (2018)
18. Esposito, C., De Santis, A., Tortora, G., Chang, H., Choo, K.K.R.: Blockchain: a panacea for healthcare cloud-based data security and privacy? IEEE Cloud Comput. **5**(1), 31–37 (2018)

19. Zhang, P., Walker, M.A., White, J., Schmidt, D.C., Lenz, G.: Metrics for assessing blockchain-based healthcare decentralized apps. In: 2017 IEEE 19th International Conference on e-Health Networking, Applications and Services (Healthcom), pp. 1–4, October 2017
20. Mettler, M.: Blockchain technology in healthcare: the revolution starts here. In: 2016 IEEE 18th International Conference on e-Health Networking, Applications and Services (Healthcom), pp. 1–3, September 2016
21. Liu, W., Zhu, S., Mundie, T., Krieger, U.: Advanced blockchain architecture for e-health systems. In: 2017 IEEE 19th International Conference on e-Health Networking, Applications and Services (Healthcom), pp. 1–6. IEEE (2017)
22. Damiani, J.: Simplyvital health is using blockchain to revolutionize healthcare (2017). https://www.forbes.com/sites/jessedamiani/2017/11/06/simplyvital-health-blockchainevolutionize-healthcare/#3777e022880a
23. Azaria, A., Ekblaw, A., Vieira, T., Lippman, A.: MedRec: using blockchain for medical data access and permission management. In: 2016 2nd International Conference on Open and Big Data (OBD), pp. 25–30, August 2016
24. Rabah, K.: Challenges & opportunities for blockchain powered healthcare systems: a review. Mara Res. J. Med. Health Sci. 1(1), 45–52 (2017). ISSN: 2523-5680. https://medicine.mrjournals.org/index.php/medicine/article/view/6
25. Kuo, T.-T., Kim, H.-E., Ohno-Machado, L.: Blockchain distributed ledger technologies for biomedical and health care applications. J. Am. Med. Inform. Assoc. JAMIA 24(6), 1211–1220 (2017). https://doi.org/10.1093/jamia/ocx068
26. Mangosalad/mercantis-distributedhealth-2017-winner (2018). https://github.com/MangoSalad/Mercantis-DistributedHealth-2017-Winner
27. Dufel, M.: A new paradigm for health information exchange. Technical report, Peer Health (2016). http://peerhealth.io/download-whitepaper/
28. Zhang, P., White, J., Schmidt, D.C., Lenz, G., Rosenbloom, S.T.: FHIRChain: applying blockchain to securely and scalably share clinical data. Comput. Struct. Biotechnol. J. 16, 267–278 (2018)
29. Archetype Designer. https://tools.openehr.org/designer/. Accessed 10 Nov 2021
30. Medblocks UI VsCode Extension. https://github.com/medblocks/vscode-medblocks-ui. Accessed 10 Nov 2021
31. openEHR Toolkit. https://toolkit.cabolabs.com/. Accessed 10 Nov 2021

Skin Cancer Recognition
for Low Resolution Images

Michał Kortała[1]([⊠])(iD), Tatiana Jaworska[2]([⊠])(iD), Maria Ganzha[1,2]([⊠])(iD),
and Marcin Paprzycki[2]([⊠])(iD)

[1] Warsaw University of Technology, Warsaw, Poland
kortalam@student.mini.pw.edu.pl, {maria.ganzha}@ibspan.waw.pl
[2] Systems Research Institute Polish Academy of Sciences, Warsaw, Poland
{tatiana.jaworska,marcin.paprzycki}@ibspan.waw.pl

Abstract. Substantial body of work has been devoted to skin cancer recognition in high-resolution medical images. However, nowadays photos of skin lesions can be taken by mobile phones, where quality of image is reduced. The aim of this contribution is report on skin cancer recognition when applying machine learning to "low resolution" images. Experiments have been performed on the dataset from the ISIC 2018 Challenge.

Keywords: Skin cancer · Machine Learning · Deep Neural Network

1 Introduction

Number of cases of skin cancer is raising. In 2017, in Poland, 7,000 new cases among women and 6,453 among men were detected, an increase of 9% compared to 2016 [36]. Currently, skin cancer diagnosis involves: (a) visual examination, and (b) skin biopsy. Too often, patients see the doctor in later stages of cancer, while early detection could ensure 5-year survival (with certainty of 99%). Beginning of treatment requires early cancer awareness. One possible solution would be a simple diagnosis tool, which would automate initial diagnosis.

With progress in image processing it should be possible to facilitate early skin cancer detection. Organizations, such as the International Skin Imaging Collaboration (ISIC), actively seek solutions by organizing competitions in this area, and engaging the research community. For instance, the ISIC 2018 Challenge [12, 41] included diagnosis of skin cancer on the basis of image analysis.

This work proposes a diagnosis system for devices with restricted resources, such as mobile phones. This is achieved by reducing the resolution of photos, rescaled to size 100×100. This allows to speed the model training and reduces the size of the model achieved (e.g. to fit the memory of standard phones).

2 Related Work

2.1 Image Preprocessing

Before applying machine learning (ML) images have to be suitably prepared. Numerous image preprocessing techniques have been explored. Farooq et al. [16]

S. Sachdeva et al. (Eds.): BDA 2021, LNCS 13167, pp. 122–148, 2022.
https://doi.org/10.1007/978-3-030-96600-3_10

applied hair removal ([26]) to clean the image. Color space transformation were used to separate color from luminance. Here, Sarkar et al. [34] proposed transformation to CIEL*a*b, Abbas et al. [1] to CIECAM02, and Kumar et al. [24] to HSV and to YCbCr color spaces. Sarkar et al. [34] used color space transformation, together with luminescence channel enhancement, applying CLAHE algorithm. Preprocessing reported in Kumar et al. [24] explored feature extraction methods, e.g. conversion to grayscale, binary mask, sharpening filter, smooth filter, adjust histograms, median filter, RGB extraction and Sobel operator.

In computer vision, a region of interest (ROI) is a part of an image identified for a particular purpose. Abbas et al. [1], used region of interest selection, and structural feature extraction, with steerable pyramid transform. An approach presented in [25], used a threshold and statistical region merging (SRM) to generate a binary mask. Moreover, Karhunen-Loeve transform, histogram equalization, and contrast enhancement were applied. Sah et al. [33] proposed image augmentation via rotation, shift, and reflection. The preprocessing step often involves convolutional neural network (CNN; see, [3]). The role of CNN is to classify input images based on features extracted in a training process. In Sedigh et al. [35], dataset augmentation, by image generation, used a generative adversarial network, to solve the problem of insufficient number of images in the training set, by creating synthetic samples. Other preprocessing methods included grey image scale, noise filtering, principal component analysis, gray level co-occurrence matrix [4], border detection [37] and color normalization [7,8].

2.2 Classifiers

After extraction of features from preprocessed images, classification ensues. Typically it involves CNNs, sometimes with transfer learning. Following approaches have been reported (for details, readers are directed to the references):

- Inception-v3 [15,16,31,40]
- InceptionResNetv2 [2,31,39]
- ResNet152 [2,18,31]
- DenseNet201 [8,21,31]
- MobileNet [16,20]
- AlexNet [10,23]
- VGG16 [33,43]
- Custom architectures [22,35].

2.3 Ensembles

Models described in Sect. 2.2 were often grouped into an ensemble of models, aimed at enhancing prediction of its members. Numerous publications suggested using one of the following approaches for building hierarchical classifiers. The first solution consists of a neural network trained on each model's output. In the second one [22,28] voting was used. Sun et al. [38] proposed an approach consisting of a dedicated ensemble for each class predicted by a classifier. Summary of pertinent models is shown in Table 1.

Table 1. Summary of proposed solutions in the area of skin cancer classification.

Authors	Summary
Farooq et al. [16]	Inception-v3, MobileNet with transfer learning
Habib et al. [31]	CNN with transfer learning
Esteva et al. [15]	Inception-v3 with transfer learning
Sarkar et al. [34]	ResNet, CNN
Abbas et al. [1]	AdaBoost
Kumar et al. [24]	Neural network, k-Near Neighbours, decision trees
Sah et al. [33]	VGG16 with transfer learning
Capdehurat et al. [11]	AdaBoost
Ahmad et al. [2]	CNN with triplet loss
Lau et al. [25]	Neural network, auto-associative neural network
Masood et al. [29]	SVM with incremental use of unlabeled data
Sedigh et al. [35]	Custom built CNN
Alquran et al. [4]	SVM
Suganya [37]	SVM
Barata et al. [8]	DenseNet-161 with transfer learning
Dorj et al. [14]	AlexNet with transfer learning, ECOC SVM

2.4 Datasets

There exist multiple datasets for skin cancer classification. They include from 2 to 10 classes. Most of them are small, e.g. DERMQUEST (76 images), PH2 (200), or Dermweb (320). However, Dermnet has 5500 images, while the dataset for the 2018 ISIC challenge consisted of 10,015 images.

2.5 Results Reported in Key Publications

Results found in relevant publications have been obtained for different datasets. This makes it difficult to fairly compare prediction quality. Nevertheless, reported results can be presented collectively (in Table 2) to illustrate key trends.

Publications with best results (above 90% of accuracy) were obtained on relatively small datasets; e.g. Sarkar et al. ([34]) used a 700 image subset of the ISIC challenge dataset. Apart from models proposed in [15] and [12], all results were trained and tested on datasets smaller than the ISIC 2018 dataset. The dataset from the ISIC Challenge 2018 is described in [12]. Here, the training set consists of 10,015 images, while test set of 1,512 images (see, also Sect. 3). The best reported result, *in terms of balanced accuracy*, was 88.5%.

2.6 Existing Software

Currently multiple software solutions are available, e.g. neural network libraries – Tensorflow, Pytorch, and Matlab toolbox. In the preprocessing stage, OpenCV implementation is often used, and the scikit-image Python package. Cited references used Tensorflow for neural networks [2,8,33] and Matlab [25,37]. In some works reported in Table 1 a software source was not specified [4,11,29,35].

Table 2. Best results reported in listed contributions.

Authors	Dataset size	Accuracy	Sensitivity	Specificity	F1	ROC AUC
Farooq et al. [16]	2637	86 %	89 %	83 %	84 %	–
Habib et al. [31]	10,015	–	–	–	89.01 %	0.987
Esteva et al. [15]	129,450	–	–	–	–	0.910
Sarkar et al. [34]	700	97.86 %	–	–	–	–
Abbas et al. [1]	1039	–	89.28 %	93.75 %	–	0.986
Kumar et al. [24]	N/A	95 %	–	–	–	–
Sah et al. [33]	5,500	82 %	83 %	82 %	–	–
Capdehurat et al. [11]	655	–	90 %	85 %	–	0.937
Ahmad et al. [2]	6,144	87.42 %	97.04 %	96.48 %	–	–
Lau et al. [25]	N/A	73 %	–	–	–	–
Masood et al. [29]	3,800	94 %	–	–	–	–
Sedigh et al. [35]	97	71 %	68 %	74 %	70 %	–
Alquran et al. [4]	N/A	92.1 %	–	–	–	–
Suganya [37]	320	96.8 %	95.4 %	89.3 %	–	–
Barata et al. [8]	2,000	70 %	–	–	–	0.876
Dorj et al. [14]	3,753	94.2 %	87.83 %	90.74 %	–	–
Codella et al. [12]	10,015	88.5 %	–	–	–	–

3 Dataset Description

Results reported here concern multi-class classification, using data available in Task 3 of the ISIC 2018 Challenge competition [12,41]. This dataset includes dermatoscopic photos, from different populations, obtained and stored in different modalities. The training part of the dataset contains 10,015 images. Available photos present a representative collection of dermatologically relevant categories of pigmented skin lesions. Categories present in the dataset include:

1. Melanoma (mel)
2. Melanocytic nevus (nv)
3. Basal cell carcinoma (bcc)
4. Actinic keratosis/Bowen's disease (akiec)
5. Benign keratosis (bkl)
6. Dermatofibroma (df)
7. Vascular lesion (vasc).

Here, Basal cell carcinoma and melanoma are severe skin cancers. Actinic keratosis is the most common form of precancer that develops on skin damaged by ultraviolet rays, and may evolve into Bowen's disease. Melanocytic nevus (common mole), is a benign skin lesion that is visually similar to (and difficult to distinguish from) melanoma. Seborrheic keratosis is a noncancerous skin lesion that some people develop as they age. Dermatofibroma is also a benign skin lesion. Majority of vascular lesions are benign. However, rarely, malignant variants can materialize. Sample images of each class, are shown in Fig. 1.

Existence of seven different classes makes the problem a multi-class classification problem. This, limits the number of possible approaches, as not every "standard classifier" supports multi-class classification.

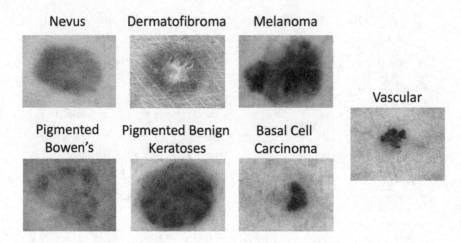

Fig. 1. Example of each database class.

Another feature of the dataset is extreme class imbalance. The largest category is more than 58 times larger than the smallest one, and more than 6 times larger than the second largest. This imbalance is illustrated in Fig. 2. This lack of balance adversely affects majority of classifiers, as they try to minimize the global error. Hence, the classifiers treat the minority classes as a "noise".

In this work, images have been resized to 100×100, from the original size of 450×600. In doing so, the images suffered some loss of details and a dimensional disturbance. However, would enable using such dataset on machines with restricted resources, and in a reasonable time. Moreover, the cost of collecting and storing data would be lowered. Finally, it would support the use of data acquired from devices from different manufacturers, which vary in size and proportions.

4   Image Preprocessing

4.1   Hair Removal

In the dataset, some skin lesions are partially covered by body hair, which can be a problem in the feature extraction and the region of interest segmentation, due to obscuring of essential features of the lesion. In order to eliminate this problem, a hair removal algorithm was developed, inspired by [26]. This algorithm was

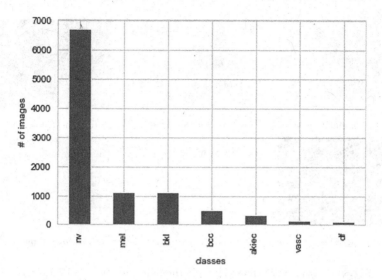

Fig. 2. Images numerosity in the dataset.

chosen, instead of one proposed in Lee et al. [26], because of its speed. The method consists of:

1. Grayscale image transformation.
2. Performing blackhat morphological operation.
3. Creating a mask by applying thresholding.
4. Inpainting mask pixels, in the original image, using the region neighborhood.

Here, when creating mask (applying thresholding), threshold value of 10 was used (every pixel with a value less than 10 was considered a skin pixel, rather than hair pixel). Pixels were filled starting at the region boundary, with the Fast Marching Method, to select the next pixel to fill the region. When hair is blonde and the skin lesion underneath it is dark, a low value of the threshold ensures that the hair will be removed, although small cavities may occur in the lesion. These will be fixed by refilling, on the basis of their neighborhood. Result of each step are shown in Fig. 3.

Proposed algorithm successfully removes hair from skin lesion images, especially thick ones, which have the worst impact on lesion visibility. All of lesion surface is clearly visible, and none of its regions are hidden behind hair.

4.2   Color Preprocessing

The images in the dataset have been acquired from multiple sources and "differ in appearance", among others, due to the light conditions. Leveling light conditions across the dataset improves the stability of classification systems [9]. To address this problem, the Shades of Gray color constancy algorithm [17], was applied.

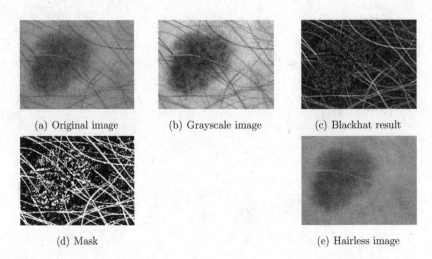

(a) Original image   (b) Grayscale image   (c) Blackhat result

(d) Mask   (e) Hairless image

Fig. 3. Illustration of main steps of the hair detection and removal.

It is to adjust image colors to look like colors under a "standard illumination". The applied approach consists of two phases:

1. Color estimation of the illuminant in the RGB color space.
2. Transformation of image using the estimated light source.

Beneath, Eq. 1 was used to estimate the color of a standard light source.

$$\left(\frac{\int I_c(\mathbf{x})^p \, d\mathbf{x}}{\int d\mathbf{x}}\right)^{\frac{1}{p}} = K e_c. \tag{1}$$

where:

- I_c stands for c-th image channel of an image I,
- $\mathbf{x} = (i, j)$ is the point describing the position of a pixel,
- K is a normalization constant, which guarantees that $\mathbf{e} = [e_R \; e_G \; e_B]^T$ vector is a unit vector,
- p is a parameter of the Minkowski norm.

In the implementation, Minkowski norm with $p = 6$ was chosen, as it has delivered the best results in [9,17]. After calculating the $\mathbf{e}$ vector, the image can be transformed according to Eq. 2, which is applied to each pixel.

$$\begin{pmatrix} I_R^* \\ I_G^* \\ I_B^* \end{pmatrix} = \begin{pmatrix} \frac{1}{\sqrt{3}e_R} & 0 & 0 \\ 0 & \frac{1}{\sqrt{3}e_G} & 0 \\ 0 & 0 & \frac{1}{\sqrt{3}e_B} \end{pmatrix} \begin{pmatrix} I_R \\ I_G \\ I_B \end{pmatrix} \tag{2}$$

Example results of performing color constancy algorithm on four images of melanocytic nevus skin cancer class are presented in Fig. 4.

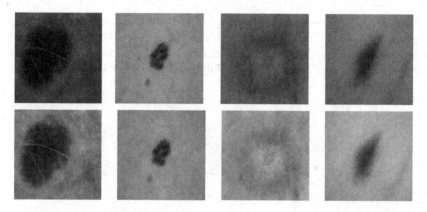

Fig. 4. Color constancy applied to melanocytic nevus images. First row contains original images, second row contains images after transformation.

In Fig. 4, initially, skin color varied greatly between images. After the transformation, there is almost no difference in perception of similar colors; modified images look "similar", independently of the original light conditions.

4.3 Histogram Equalization

Even after two steps of preprocessing, not every skin lesion had a clear border, or clearly visible details. Hence, the Contrast Limited Adaptive Histogram Equalization (CLAHE) algorithm was applied. The CLAHE differs from classic histogram equalization algorithms in two key aspects:

- It operates on neighborhood regions, instead of the full image.
- It limits the intensification, by trimming the histogram at the preset value and redistributing that part of the histogram uniformly across all bins.

Equalizing histogram in regions, instead of the full image, tends to work better on images with regions that are considerably darker, or lighter, than the rest of the image, as it provides better enhancement of contrast. However, working on regions could over-amplify the contrast in near-constant regions, because the histogram is highly concentrated in such regions. Hence, contrast limiting is used. It clips the pixels from bins that exceed the limit value and redistributes that part of the histogram uniformly across other bins.

While original images are encoded in RGB color space, application of CLAHE requires using a color space with a brightness channel. Applying CLAHE to each RGB channel can introduce significant loss in color information. Hence, images were converted to one of the following color spaces:

- HSV
- YCrCb
- CIEL*a*b.

Next, CLAHE was applied to the V, Y and L channels, before converting the image back to the RGB color space. This better preserved the color balance of the image, as above color spaces have channels that separate color information from their brightness. The results of applying CLAHE are shown in Fig. 5.

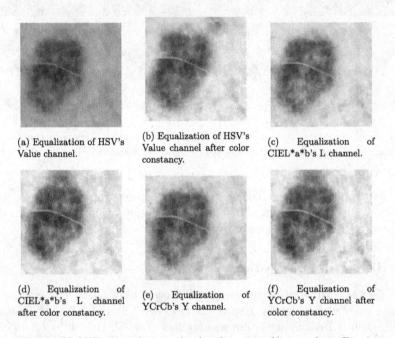

(a) Equalization of HSV's Value channel.

(b) Equalization of HSV's Value channel after color constancy.

(c)   Equalization   of CIEL*a*b's L channel.

(d)   Equalization   of CIEL*a*b's   L   channel after color constancy.

(e)   Equalization   of YCrCb's Y channel.

(f)   Equalization   of YCrCb's Y channel after color constancy.

Fig. 5. CLAHE algorithm results for the original image from Fig. 4

Upon examination it is clear that the enhanced images have better contrast, resulting in better separation between skin lesion and the regular skin area.

4.4   Region of Interest

In image processing, region of interest (ROI) is defined by a binary mask (same size as in the original image) with pixels having value 1 if they belong to ROI, and 0 otherwise. In skin lesions analysis, ROI masks indicate location (continuous region) of the (primary) skin lesion. For locating ROI, solution presented in [5,6] was used. It is a deep auto-encoder-decoder, extending the U-net neural network with bidirectional convolutional LSTM layers, to encode both semantic and high-resolution information. After establishing the mask, the region of interest was extracted from original images, as depicted in Fig. 6.

5   Feature Extraction

Preprocessed images were used as an input to feature extractors, such as Convolutional Neural Networks (CNN; [30]) and Histogram of Oriented Gradients

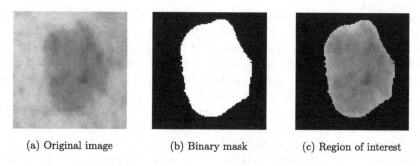

(a) Original image   (b) Binary mask   (c) Region of interest

Fig. 6. Region of interest extraction steps.

(HOG; [13]). This allows comparing quality of features obtained by each of them, and measuring potential improvement resulting from their use.

5.1   CNNs as Feature Extractors

Modern feature extractors tend to use not only convolutional layers of various sizes but also multiple forms of connection – serial and parallel. Some of these extractors introduce other advanced types of connections, e.g. residual and bidirectional ones. The following network architectures were considered in this work:

- ResNet152V2 [18]
- InceptionV3 [40]
- MobileNet [20]
- DenseNet169 [21]
- InceptionResNetV2 [39]
- VGG16 [43].

A common problem in deep learning is the *vanishing gradient*. It occurs when, during gradient propagation, in the training process, consecutive multiplications result in smaller numbers. As a result, the calculated gradient diminishes almost to zero, stopping weights from improving, or drastically slowing the process. The ResNet152V2 architecture addresses this problem. It introduces a residual ("shortcut") connections between subsequent layers, which skip at least one layer in the network. Here, a network block, instead of learning an underlying function $\mathcal{G}(x)$, learns the residual $\mathcal{F}(x) = \mathcal{G}(x) - x$ where x denotes the input of a block. Such shortcut paths "carry the gradient" throughout the length of a network.

Inception network addresses the problem of extensive networks. Instead of elongating the network, it makes it "wider", and introduces parallel layers. The building block is an inception module. Each block extracts multiple types of features – globally extracted by 5×5 convolution, local features extracted by 3×3, and local features that distinguish themselves from their neighborhood.

MobileNet takes advantage of depthwise separable convolution. In the first step, a 2D convolutional filter is applied to each channel, separately. Next, resulting feature maps are stacked, and 1×1 convolutional filter is applied across channels. Depthwise separable convolution limits the number parameters and multiplications. This makes MobileNet faster and less prone to overfitting.

DenseNet, similarly to ResNet, reduces a vanishing gradient problem by concatenating outputs of each convolutional block with outputs of a preset number of previous convolutional blocks inside a dense block. Overall, DenseNet does not sum output feature maps, but concatenates them.

InceptionResNet architectures combine best features from Inception networks and ResNets. It extends regular inception block with a "shortcut connection" between an input and an output of a block, as in the residual block in ResNets.

VGG16 consists of 16 layers that aim to reduce the number of parameters by replacing larger convolutional kernels with multiple 3 × 3 ones. For example, a 5 × 5 kernel could be replaced by two 3 × 3 kernels, reducing the number of parameters from 25 to 18 (28% drop in size).

These architectures are used primarily in two ways. (1) They are trained from scratch, to extract features present in given dataset. (2) They are used jointly with transfer learning in which weights are ultimately "frozen". The networks with frozen weights are treated as external feature extractors. Typically, ImageNet [32] dataset was used for transfer learning. This approach was proven to be a good choice in [33], and in multiple ISIC 2018 competition submissions.

5.2 Histogram of Oriented Gradients

Another popular feature extractor is the histogram of oriented gradients (HOG) method [13]. Here, it is assumed that local object structure, and appearance, can be described by the distribution of intensity of oriented gradients. The algorithm consists of following four steps. (1) Calculation of horizontal and vertical gradients. To do so, the image is filtered with kernels presented in Eq. 3. These kernels extract edges in horizontal and vertical directions. On the basis of this information gradient is computed.

$$(-1\ 0\ 1)\ \begin{pmatrix} -1 \\ 0 \\ 1 \end{pmatrix} \tag{3}$$

(2) The image is divided into cells with preset size. In each cell, histogram of gradients is calculated. (3) Cells are grouped into larger blocks and gradient strength is normalized inside the block. The block normalization is done by using the L2-hys norm, which is equal to the standard L2-norm, followed by clipping the vector values to the maximum value of 0.2 and renormalizing it. Equation 4 presents the normalization factor f of L2-norm.

$$f = \frac{v}{\sqrt{\|v\|_2^2 + e^2}} \tag{4}$$

where:

- v denotes the vector,
- e stands for a small constant value.

(4) The final feature vector is built by concatenating vectors obtained from each block. In the presented work, cells of size 4×4 are used, with 8×8 blocks, as they proved to achieve best results. The visualization of HOG feature descriptors is shown in Fig. 7.

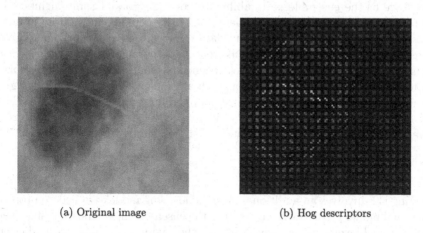

(a) Original image   (b) Hog descriptors

Fig. 7. Visualization of HOG feature descriptors.

As can be seen, HOG detects a skin lesion on the image that is represented as a mask with marked feature descriptors. One can see the shape of the considered skin lesion with the naked eye.

6 Classification Models

In order to decide which type of skin lesion a given photo contains, basis of the literature review, the following machine learning models were experimented with:

- Support Vector Machine,
- Decision Tree,
- Boosting(AdaBoost/XGBoost),
- Random Forrest,
- Fully Connected Artificial Neural Networks,
- Ensemble of experts.

All models were also considered for ensemble learning, described in what follows.

7   Ensemble Learning

Ensemble learning is a process of using multiple machine learning models to improve their individual performance. The ensemble methods usually improve results when there is significant diversity among the member models. The main advantage of the ensemble is its ability to correct errors of some members on the basis of "opinions of other members". Diversity could be introduced in many ways, e.g. training models on different datasets or subsets of the original dataset or using different models and/or different hyperparameters. Ensemble learning is used to combine multiple heterogeneous models, where each performs full multiclass classification. This approach aims at polishing the already "good enough" classification of members, instead of training them from scratch. Here, only lightweight approaches were considered, as the proposed system aims at being used on machines with restricted resources, similar to that used in [27].

7.1   Stacking

Stacking [42] involves an additional machine learning model, trained "on top" of individual model predictions. In our case, the ensemble members are trained separately, to achieve best accuracy. Next, another model is trained, on predictions of ensemble members to deliver the final prediction.

7.2   Voting

Voting is the simplest way to create an ensemble. It acts solely on labels and does not attach importance to the source of prediction. As its result, voting ensemble chooses the label, which gets most votes. The voting process could be executed in two ways as majority voting or weighted majority voting. In majority voting all votes of members of an ensemble have the same weight.

In weighted majority voting, each vote is scaled taking into account perceived "strength" of individual models and "boost" these that perform better. While any metric can be used to generate weights, in what follows, a balanced accuracy metric was used.

8   Experimental Results

In this section the results of a comprehensive set of experiments are presented and analysed. We start from the description of the experimental setup.

8.1   Experimental Setup

Performance of all models has been evaluated against the original goal metric of the ISIC 2018 Challenge, the *balanced accuracy*, which is defined as an average of recalls obtained on each class along with standard accuracy. Recall performance of classification of each class is represented in the form of confusion matrix. For all these metrics, standard definitions are applied (see, for instance [19]).

Each reported method has been implemented using Python, with Tensorflow, OpenCV, scikit-learn and the Python Imaging Library.

Classifiers were tested on two datasets. The first one internally separated from the competition training dataset. This dataset was split into train (80%) and test (20%) sets, ensuring that both sets contain representatives of all classes. The second dataset was the official dataset of competition. Models were first evaluated on the internal test set and the ones with promising results were then checked against the official one. In all the experiments, the same hair removal algorithm has been applied (see, Sect. 4.1). Detailed parameters of each model are presented in subsequent sections, related to each experiment.

8.2   HOG-Based Feature Extraction

The first set of experiments concerned the HOG algorithm, used as a feature extractor. Before extracting features, the color constancy algorithm was used. To compensate for the disproportion between the number of images in each class, simple oversampling has been applied. The images from classes with smaller sizes have been duplicated, to roughly match the number in the largest class. Next, random horizontal and vertical rotation were applied. HOG parameters were selected as described in Sect. 5.2. The results obtained on the internal dataset are presented in Table 3.

Table 3. Classification results; internal test set; HOG as a feature extractor.

Model	Accuracy	Balanced accuracy
ANN	67.09	41.25
Decision Tree	78.44	14.28
AdaBoost	78.44	14.28
XGBoost	81.70	14.28
Random Forest	2.53	14.28
SVM	58.68	34.79

The features generated using HOG contain mostly shape and geometric information. This proved to be insufficient for tree-based methods, as they classified images as belonging to the same category. In contrast, SVM and ANN achieved far better results. Note that, with limited size of the images, these models managed to outperform some of results submitted to the ISIC 2018 Challenge.

8.3   CNNs as Feature Extractors for Tree-Based Classifiers

In subsequent experiments, convolutional neural networks were used as feature extractors. Overall, for tree-based classifiers CNNs were much better feature

Table 4. Configuration of tree-based models.

Model	max depth	max leaf nodes	# estimators	lr	alpha
Decision Tree	5	200	–	–	–
AdaBoost	5	200	50	0.005	–
XGBoost	5	200	50	0.5	0.3
Random Forest	5	200	100	–	–

extractors than HOG. This is because they are able to extract more complex properties from an image. Their configuration, used in the experiments is summarized in Table 4. Both boosting algorithms used the same number of estimators inside. In each classifier a tree was limited to a depth level equal to 5 and a maximum number of leaf nodes equal to 200.

Before feature extraction, images have been processed using color constancy algorithm (best for tree-based methods). Classifiers were evaluated on features extracted by convolutional parts of ResNet152V2, DenseNet169 and MobileNet networks. Each CNN was used as an external extractor with weights obtained from training it on the ImageNet. We decided to use it despite of transfer learning rules which assume that the distribution of both sets (training and destinating) should have the same distribution.

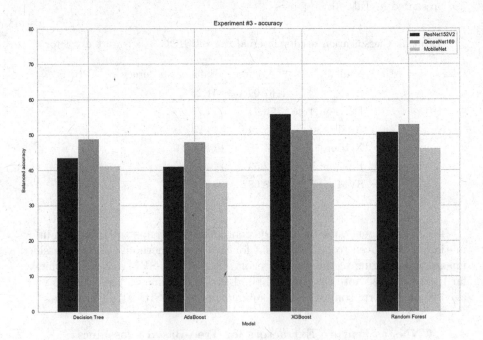

Fig. 8. Comparison of results achieved on the tree based model.

Table 5. Best result of tree-based classifiers.

Model	Accuracy	Balanced accuracy
Decision Tree	53.40	48.89
AdaBoost	67.09	47.93
XGBoost	82.86	55.86
Random Forest	71.12	52.92

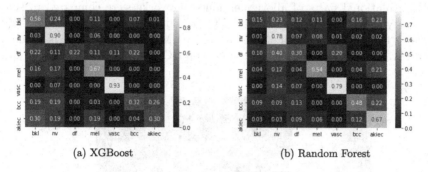

     (a) XGBoost   (b) Random Forest

Fig. 9. Confusion matrices of the best tree-based solutions.

Direct comparison of feature extractors is shown in Fig. 8 along with the best result for each classifier presented in Table 5. In Fig. 9 confusion matrices of XGBoost and Random Forest were presented as these models achieved the best results (balanced accuracy) equal to 55.86% and 52.92%, respectively.

As can be seen from confusion matrices, XGBoost achieved high degree of recognition of images belonging to the vascular lesions and the melanoma classes, above 65% of recall. Random Forest recognized around 50% of skin lesions belonging to the most dangerous classes – melanoma and basal cell carcinoma. However, these two models made mistakes for other classes, e.g. actinic keratosis was classified with approximately 30% accuracy. These two classifiers outperformed the remaining tree-based ones by up to 7% points. The overall performance of this class of models, trained with CNN extracted features, was unquestionably better than the ones using HOG as feature extractor.

8.4   SVM Classifier

The SVM classifier has been tested similarly to the tree-based models. The classification model was run using RBF kernel, with $\gamma = \frac{1}{nx_{var}}$, where n denotes number of features in the training set and x_{var} stands for variance of elements in that set. Regularization parameter was set to 1. SVM has been trained on features provided by ResNet152V2, InceptionV3 and VGG16 networks. Metrics obtained on the internal test set are shown in Table 6.

Table 6. Results of SVM with CNNs.

Feature extractor	Accuracy	Balanced accuracy
ResNet152V2	56.48	36.48
IncpetionV3	60.11	50.04
VGG16	74.34	55.09

Convolutional parts of above-mentioned networks were compared in terms of quality of features they produced. VGG16 appears to produce features that are best for the SVM. With these features, SVM achieved 55.09% of balanced accuracy, improving upon result produced on HOG features by more than 37%.

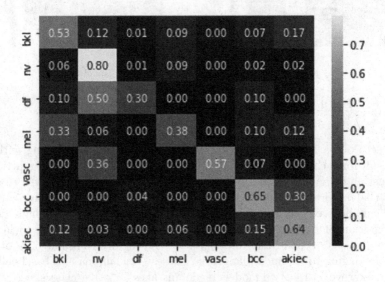

Fig. 10. Confusion matrix of SVM with VGG16.

In confusion matrix, shown in Fig. 10, one can see that SVM has a much better result in diagnosing basal cell carcinoma, as it recognized 65% of all instances present in the dataset.

8.5   Comparison of Imbalance Handling Methods

Due to the existing imbalance in the dataset, correct classification becomes a challenge. Hence, sample weighting and oversampling with random rotations were used. These methods have been evaluated on a fully connected ANN, built on the top of the ResNet152V2 feature extractor. Each time, the model has been trained for 100 epochs with a batch size of 96 images with categorical crossentropy as the loss function. The top-level network consisted of 3 hidden layers, each having 4096 neurons, with a sigmoid activation function and a dropout set equal to 0.5. The sample weights for sample s_C belonging to a class C were calculated as follows:

$$w_{s_C} = \frac{n}{N \cdot |C|} \qquad (5)$$

where N – number of classes, n – number of samples in the dataset, $|C|$ – number of samples in class C. In Fig. 11 confusion matrices corresponding to the results in Table 7 are shown. The results obtained with oversampling show certain improvement in contrast to using sample weights to counter the imbalance of the training data set. Models trained with oversampling tend to recognize either more classes, or have better average class recognition.

Table 7. Balanced accuracy of ANN models with different imbalance handling.

Imbalance handling	ImageNet	Random weights
Oversampling	65.29	42.14
Sample weight	62.43	41.29

Table 7 shows that using oversampling is preferred over the application of sample weights. Results achieved on the augmented dataset allowed to improve the balanced accuracy of the model with transfer learning by 2.86% and of the one trained from scratch by 0.85%.

As can be seen in Figs. 11 and 12 using oversampling increases the lowest nonzero recall by up to 10% points, while increasing the average value of this metric for all classes.

8.6   Region of Interest

Another experiment was performed using the ROI extraction described in Sect. 4.4. Separately, convolutional parts of networks described in Sect. 5.1 were used as feature extractors. The result of an experiment comparing preprocessed and full images is presented in Table 8. Here, top-level architecture consisted of three hidden layers with 4096 neurons, each with a sigmoid activation function and dropout of 0.5.

Table 8. Balanced accuracy of ANN models trained on ROIs and full images.

Imbalance handling	ROI	Full image
ResNet152V2	26.43	43.86
InceptionV3	36.00	46.57
MobileNet	26.42	40.14
DenseNet169	45.57	43.85
InceptionResNetV2	36.01	41.86
VGG16	14.29	14.29

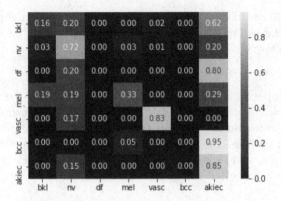

(a) From scratch with sample weight

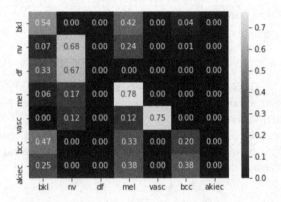

(b) From scratch with oversampling

Fig. 11. Confusion matrices of networks trained from scratch.

Extracting ROIs tends to perform worse than networks trained with exactly the same parameters and applied to full images. As can be seen (in Table 8) only DenseNet169 improves its predictions when trained on ROIs. This effect is due to the large number of black pixels introduced into the photo during the region of interest extraction. Other architectures do not seem to handle this properly as it likely results in "deactivation" of a big number of neurons.

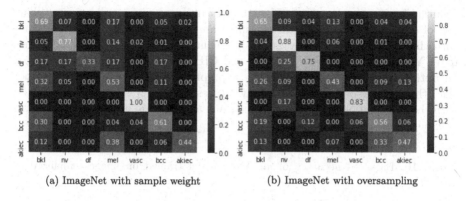

(a) ImageNet with sample weight   (b) ImageNet with oversampling

Fig. 12. Confusion matrices of networks with transfer learning

8.7 Pretrained Networks Vs Learning from Scratch

Overall, the dataset consists of insufficient number images for CNN training from scratch. Even after oversampling with augmentation, the size of the dataset is equal to 23,450 records. To overcome this problem, pretrained on the ImageNet dataset weights were used. The primary motivation was to utilize low-level features that networks learned to recognize. In the learning process of a top-level dense network has been trained to build upon those features. This is also why time needed to complete training on the dataset of interest is shortened.

This approach proved to succeed in the original ISIC 2018 competition, though when applied to bigger images. In the summary of results presented in the Table 9 models that achieved below 30% of balanced accuracy in at least one category are omitted. One can see that ResNet152V2 performed extremely well using the weights trained on the ImageNet. It improved its result by 21.43% points while keeping the training duration the same.

Table 9. Balanced accuracy ANN models with respect to the CNN part and initial weights.

CNN model	ImageNet	Random weights
ResNet152V2	65.29	43.86
MobileNet	36.60	40.14
DenseNet169	51.56	43.85

8.8 Histogram Equalization

The next set of experiments was aimed at verification of usefulness of histogram equalization methods proposed in Sect. 4.3. These methods were evaluated to check, which approach is better to handle different light conditions for images acquired from multiple sources. Here, impact of histogram equalization on channels of different color spaces was measured and compared to the result obtained with the use of the color constancy algorithm. CLAHE algorithm was applied to V, Y, L channels of HSV, YCrCb and CIEL*a*b color spaces, respectively, for both pure RGB images and for those preprocessed with the color constancy algorithm. The experiment was done with ResNet152V2 as a feature extractor, with the top-level dense layers with the same architecture as previously. Results of experiments are summarized in Table 10.

Table 10. Balanced accuracy of an ANN model trained on images enhanced by CLAHE.

Enhancement	Balanced accuracy
Color constancy	65.29
HSV color constancy	57.76
YCrCb color constancy	55.12
CIEL*a*b color constancy	57.23
HSV	48.22
YCrCb	46.67
CIEL*a*b	56.42

One can see that the color constancy algorithm achieved the best result as it normalized colors present in the image and made them more comparable. Histogram equalization done by CLAHE on regular RGB images works a little bit worse. It noticeably increases brightness level but does not impact the color differences, leaving them "hard to compare". CLAHE applied to the color constancy images, as discussed in Sect. 4.3, can enhance some darker regions of the skin around the lesion, making it indistinguishable from the lesion and therefore confusing a classifier.

8.9   Ensemble of Experts

In this section, problem of multiclass classification has been replaced with a group of simpler problems as described in [28]. As a result, an ensemble of experts is built consisting of models responsible for predicting only one class from the dataset. Table 11 presents the accuracy and the balanced accuracy of each expert along with results from the ensemble.

As one can see, higher results achieved by individual expert in its respective task confirm that simplifying the problem influences the result. Ensemble of experts classifier achieved over 60% of balanced accuracy on the internal test set, which makes it the second and the third best solution considered in this work.

Table 11. Results of individual experts and the full ensemble.

Expert	Accuracy	Balanced accuracy
bkl	84.42	84.69
nv	91.67	85.35
df	98.91	74.82
mel	77.35	83.44
vasc	96.01	65.02
bcc	96.01	72.77
akiec	96.20	85.91
ANN	78.26	63.20
XGBoost	84.60	57.72
Voting	82.97	61.50

Predictions made by experts were combined together using one of three techniques: expert voting, dense neural network and extreme gradient boosting. Confusion matrices with results from ensembles are shown in Fig. 13. It is clearly visible that combining results of individual experts by the process of voting brings the best results. It has the highest average recall and the highest minimal recall across all models proposed here. The results may be surprising because this approach is not very popular in the literature as stated in Sect. 2.

8.10   Summary of Results for the ISIC 2018 Challenge

In Table 12 a summary of results achieved by each considered model is shown. Classifiers with balanced accuracy above 55% (on the internal dataset) were evaluated on the original ISIC 2018 Challenge test set. Italicized text denotes models trained on RGB images without color constancy.

Table 12. Results summary of top 10 results.

Extractor	Model	Internal test set	External test set
ResNet152V2	XGBoost	55.86	41.40
ResNet152V2	ANN	65.29	49.60
ResNet152V2 CIELAB	ANN	56.52	48.0
ResNet152V2 CIELAB	ANN	57.23	42.1
ResNet152V2 HSV	ANN	57.76	46.2
ResNet152V2 YCrCb	ANN	55.12	39.0
VGG16	SVM	55.09	43.1
ResNet152V2 Expert	Voting	61.50	48.9
ResNet152V2 Expert	ANN	63.20	46.6
ResNet152V2 Expert	XGBoost	57.72	40.1

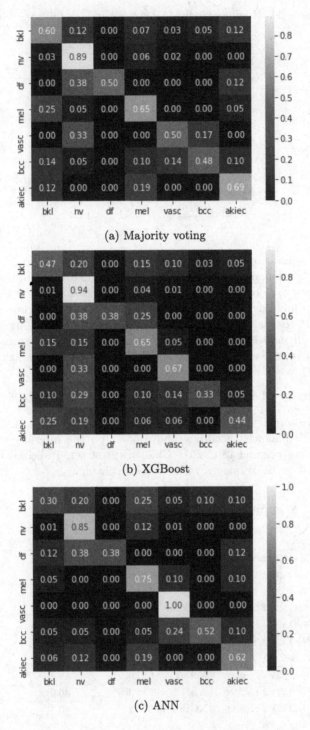

(a) Majority voting

(b) XGBoost

(c) ANN

Fig. 13. Confusion matrices of an ensemble of experts.

8.11   Results Obtained for Ensemble Learning

In this section, the comparison of results obtained with ensemble learning applied is presented. As ensemble members, models (reported above) with results above 42%, have been selected. Both versions of voting along with extreme gradient boosting and artificial neural networks, have been tested. The results of these experiments are presented in Table 13.

Table 13. Results of an ensemble learning with 7 best performing models as its members.

Type	External test set
Majority voting	52.8
Weighted majority voting	52.7
ANN stacking	45.3
XGBoost stacking	40.0

As one can see the final ensemble achieved the highest balanced accuracy, by applying voting to combine predictions from member models. The use of an ensemble of models increased the final classification by 3% points.

The results are promising in the context of computational resources available, as well as size of images. The proposed solution would take 60th place in the ISIC 2018 Challenge, ahead of 18 other solutions, including 3 using data from external datasets, thus increasing the number of photos at the training stage.

9   Concluding Remarks

The aim of the reported work was to study application of machine learning to the skin cancer detection problem. In this project the target dataset was highly unbalanced, which introduced an additional level of difficulty to the problem.

The proposed approach differs significantly from those proposed in the literature. The difference materializes, primarily, in terms of size of the images, and the diversity of models used as members in the ensemble learning. The main idea of the proposed approach was to develop a diagnosis system that can run efficiently on mobile devices and can be trained without corporate-level computational resources (e.g. large Google/Amazon/Microsoft cloud). The modular architecture of the proposed system allows it to be easily expanded, as every stage could be replaced with another approach to its implementation, with possible further gains in performance.

Presented results show that using smaller images still allows achieving similar outcomes on comparable technical setup and computational power but reduces the complexity of the training process. The result of that is a faster learning stage and quicker predictions that could be used on mobile devices, achieving almost a real-time prediction. A single training epoch of best performing models took approximately between 57 to 85 s on NVIDIA GeForce GTX 1050 Ti 4 GB.

Although the results presented in Sect. 8 may seem worse than the best reported for the ISIC 2018 Challenge, in terms of resources used and learning time, they turn out to be more usable in everyday life. As a result, they offer hope for faster universal adaptation and greater availability of high-quality automated early diagnosis in the future.

As for the prospects for future work, the introduction of smaller, more condensed feature extractor, training on a larger dataset, with possible use of unlabeled data could be considered. Due to the proposed size of images, the execution time of an individual prediction should not rise with training performed on a bigger dataset, so the aspect of use on mobile devices could be maintained. Alongside proposed, slight enlargement of an image size aimed at restoring original image proportions could be considered.

References

1. Abbas, Q., Celebi, M., Serrano, C., Fondón García, I., Ma, G.: Pattern classification of dermoscopy images: a perceptually uniform model. Pattern Recogn. **46**(1), 86–97 (2013)
2. Ahmad, B., Usama, M., Huang, C.M., Hwang, K., Hossain, M.S., Muhammad, G.: Discriminative feature learning for skin disease classification using deep convolutional neural network. IEEE Access **8**, 39025–39033 (2020)
3. ALEnezi, N.S.A.: A method of skin disease detection using image processing and machine learning. Procedia Comput. Sci. **163**, 85–92 (2019). https://doi.org/ 10.1016/j.procs.2019.12.090. http://www.sciencedirect.com/science/article/pii/ S1877050919321295. 16th Learning and Technology Conference 2019 Artificial Intelligence and Machine Learning: Embedding the Intelligence
4. Alquran, H., et al.: The melanoma skin cancer detection and classification using support vector machine. In: 2017 IEEE Jordan Conference on Applied Electrical Engineering and Computing Technologies (AEECT), pp. 1–5 (2017)
5. Asadi-Aghbolaghi, M., Azad, R., Fathy, M., Escalera, S.: Multi-level context gating of embedded collective knowledge for medical image segmentation (2020)
6. Azad, R., Asadi-Aghbolaghi, M., Fathy, M., Escalera, S.: Bi-directional ConvLSTM U-Net with Densley connected convolutions (2019)
7. Barata, C., Celebi, M.E., Marques, J.S.: Improving dermoscopy image classification using color constancy. IEEE J. Biomed. Health Inform. **19**(3), 1146–1152 (2015)
8. Barata, C., Marques, J.S.: Deep learning for skin cancer diagnosis with hierarchical architectures. In: 2019 IEEE 16th International Symposium on Biomedical Imaging (ISBI 2019), pp. 841–845 (2019)
9. Barata, C., Celebi, M.E., Marques, J.S.: Improving dermoscopy image classification using color constancy. IEEE J. Biomed. Health Inform. **19**(3), 1146–1152 (2015). https://doi.org/10.1109/JBHI.2014.2336473
10. Brinker, T., et al.: Skin cancer classification using convolutional neural networks: systematic review. J. Med. Internet Res. **20**(10), e11936 (2018)
11. Capdehourat, G., Corez, A., Bazzano, A., Alonso, R., Musé, P.: Toward a combined tool to assist dermatologists in melanoma detection from dermoscopic images of pigmented skin lesions. Pattern Recogn. Lett. **32**(16), 2187–2196 (2011)

12. Codella, N.C.F., et al.: Skin lesion analysis toward melanoma detection 2018: a challenge hosted by the international skin imaging collaboration (ISIC). CoRR abs/1902.03368 (2019). http://arxiv.org/abs/1902.03368

13. Dalal, N., Triggs, B.: Histograms of oriented gradients for human detection. In: 2005 IEEE Computer Society Conference on Computer Vision and Pattern Recognition (CVPR 2005), vol. 1, pp. 886–893 (2005). https://doi.org/10.1109/VPR.2005.177

14. Dorj, U.O., Lee, K.K., Choi, J.Y., Lee, M.: The skin cancer classification using deep convolutional neural network (report). Multimedia Tools Appl. **77**(8), 9909 (2018)

15. Esteva, A., et al.: Dermatologist-level classification of skin cancer with deep neural networks. Nature **542**(7639), 115 (2017)

16. Farooq, M.A., Khatoon, A., Varkarakis, V., Corcoran, P.: Advanced deep learning methodologies for skin cancer classification in prodromal stages (2020)

17. Finlayson, G., Trezzi, E.: Shades of gray and colour constancy, pp. 37–41, January 2004

18. He, K., Zhang, X., Ren, S., Sun, J.: Deep residual learning for image recognition (2015)

19. Hossin, M., Sulaiman, M.N.: A review on evaluation metrics for data classification evaluations. Int. J. Data Mining Knowl. Manage. Process **5**, 01–11 (2015). https://doi.org/10.5121/ijdkp.2015.5201

20. Howard, A.G., et al.: MobileNets: efficient convolutional neural networks for mobile vision applications (2017)

21. Huang, G., Liu, Z., van der Maaten, L., Weinberger, K.Q.: Densely connected convolutional networks (2018)

22. Kannojia, S.P., Jaiswal, G.: Ensemble of hybrid CNN-ELM model for image classification. In: 2018 5th International Conference on Signal Processing and Integrated Networks (SPIN), pp. 538–541 (2018)

23. Krizhevsky, A., Sutskever, I., Hinton, G.E.: ImageNet classification with deep convolutional neural networks. In: Pereira, F., Burges, C.J.C., Bottou, L., Weinberger, K.Q. (eds.) Advances in Neural Information Processing Systems, vol. 25. Curran Associates, Inc. (2012). https://proceedings.neurips.cc/paper/2012/file/c399862d3b9d6b76c8436e924a68c45b-Paper.pdf

24. Kumar, V.B., Kumar, S.S., Saboo, V.: Dermatological disease detection using image processing and machine learning. In: 2016 Third International Conference on Artificial Intelligence and Pattern Recognition (AIPR), pp. 1–6 (2016)

25. Lau, H.T., Al-Jumaily, A.: Automatically early detection of skin cancer: study based on neural network classification. In: 2009 International Conference of Soft Computing and Pattern Recognition, pp. 375–380 (2009)

26. Lee, T., Ng, V., Gallagher, R., Coldman, A., McLean, D.: Dullrazor®: a software approach to hair removal from images. Comput. Biol. Med. **27**(6), 533–543 (1997)

27. Lin, T.C., Lee, H.C.: Skin cancer dermoscopy images classification with meta data via deep learning ensemble. In: 2020 International Computer Symposium (ICS), pp. 237–241 (2020). https://doi.org/10.1109/ICS51289.2020.00055

28. Liu, W., Zhang, M., Luo, Z., Cai, Y.: An ensemble deep learning method for vehicle type classification on visual traffic surveillance sensors. IEEE Access **5**, 24417–24425 (2017)

29. Masood, A., Al-Jumaily, A.: Semi advised learning and classification algorithm for partially labeled skin cancer data analysis. In: 2017 12th International Conference on Intelligent Systems and Knowledge Engineering (ISKE), pp. 1–4 (2017)

30. O'Shea, K., Nash, R.: An introduction to convolutional neural networks. CoRR abs/1511.08458 (2015). http://arxiv.org/abs/1511.08458

31. Rezvantalab, A., Safigholi, H., Karimijeshni, S.: Dermatologist level dermoscopy skin cancer classification using different deep learning convolutional neural networks algorithms (2018)

32. Russakovsky, O., et al.: ImageNet large scale visual recognition challenge. Int. J. Comput. Vis. (IJCV) **115**(3), 211–252 (2015). https://doi.org/10.1007/s11263-015-0816-y

33. Sah, A.K., Bhusal, S., Amatya, S., Mainali, M., Shakya, S.: Dermatological diseases classification using image processing and deep neural network. In: 2019 International Conference on Computing, Communication, and Intelligent Systems (ICC-CIS), pp. 381–386 (2019)

34. Sarkar, R., Chatterjee, C.C., Hazra, A.: A novel approach for automatic diagnosis of skin carcinoma from dermoscopic images using parallel deep residual networks. In: Singh, M., Gupta, P.K., Tyagi, V., Flusser, J., Ören, T., Kashyap, R. (eds.) ICACDS 2019. CCIS, vol. 1045, pp. 83–94. Springer, Singapore (2019). https://doi.org/10.1007/978-981-13-9939-8_8

35. Sedigh, P., Sadeghian, R., Masouleh, M.T.: Generating synthetic medical images by using GAN to improve CNN performance in skin cancer classification. In: 2019 7th International Conference on Robotics and Mechatronics (ICRoM), pp. 497–502 (2019)

36. Statistics Poland, Social Surveys Department, Statistical Office in Krakow: Health and health care in 2019 (2021)

37. Suganya, R.: An automated computer aided diagnosis of skin lesions detection and classification for dermoscopy images. In: 2016 International Conference on Recent Trends in Information Technology (ICRTIT), pp. 1–5 (2016)

38. Sun, H., Yang, J.: Domain-specific image classification using ensemble learning utilizing open-domain knowledge. In: 2019 International Conference on Computing, Networking and Communications (ICNC), pp. 593–596 (2019)

39. Szegedy, C., Ioffe, S., Vanhoucke, V., Alemi, A.: Inception-v4, inception-ResNet and the impact of residual connections on learning (2016)

40. Szegedy, C., et al.: Going deeper with convolutions (2014)

41. Tschandl, P., Rosendahl, C., Kittler, H.: The HAM10000 dataset: a large collection of multi-source dermatoscopic images of common pigmented skin lesions. Sci. Data **5** (2018). https://doi.org/10.1038/sdata.2018.161

42. Wolpert, D.: Stacked generalization. Neural Netw. **5**, 241–259 (1992). https://doi.org/10.1016/S0893-6080(05)80023-1

43. Zisserman, A.: Very deep convolutional networks for large-scale image recognition. arXiv:1409.1556, September 2014

Forecasting AQI Data with IoT Enabled Indoor Air Quality Monitoring System

Mayank Deep Khare⬤, Kumar Satyam Sagar⬤, Shelly Sachdeva$^{(\boxtimes)}$⬤,
and Chandra Prakash⬤

National Institute of Technology, Delhi, New Delhi, India
{Mayankdeep,202211010,shellysachdeva,cprakash}@nitdelhi.ac.in

Abstract. Air Pollution is one of the major problems in today's world. Automobiles, factories, power plants etc., have made human life easy, but we are compromising the environment. Air pollution is one of the negative impacts that come from these developments. An indexing system has been developed for quantitative analysis of air quality, known as Air quality index. AQI's value depends on various pollutant values such as PM (Particulate matter), CO, NH_3, NO_2, H_2S etc. Based on past data of AQI, predictions can be done for future AQI values. Significant challenges encountered in AQI monitoring are accuracy of forecasted value and indoor AQI sensing nodes with power efficiency. In this paper, we developed an indoor IoT-based AQI Monitoring sensing node to get the value of the above pollutants in the environment. With results we created a data set for forecasting AQI value. For better accuracy we applied SARIMAX and got better results from other forecasting methods such as ANN and RNN.

Keywords: AQI · Particulate matter (PM) · Forecasting · Pollutant

1 Introduction and Motivation

We are living in the era of smart connected devices. Today, we all have something in our hands, an IoT device known as a smartphone [1]. The ever-growing network of technologies connecting and communicating via the internet to send and receive data in the absence of human-to-human or human-to-computer interaction is called the 'Internet of Things (IoT).' IoT consists of a network of smart devices, sensors, transducers, and actuators communicating with each other over the internet. With Industry 4.0 revolution, IoT is rapidly evolving throughout the embedded industry. It is projected that there will be about 50 billion IoT devices connected to the internet by 2030. On the other hand, Air Pollution is one of the major problems in today's world as humans tend towards comfortable life with automobiles, factories, power plants. As a consequence, it has had a negative impact on our environment. Air pollution is one of the negative impacts that come from these developments. Different breathing-related problems like asthma, COPD, lung cancer occur due to living continuously in poor air quality. Air quality index

S. Sachdeva et al. (Eds.): BDA 2021, LNCS 13167, pp. 149–158, 2022.
https://doi.org/10.1007/978-3-030-96600-3_11

(AQI) is a measurement unit by which we check the quality of air. For calculation of air quality index, we refer to values of different pollutants like particulate matter (PM) PM2.5, PM10, NO_2, CO, CO_2, NH_3 etc. Government agencies and other air quality data repositories set up pollution monitoring nodes and regularly check the air quality index. The higher the AQI, the higher the air pollution level, raising health concerns. Various research included in the literature survey has shown different methodologies to monitor air quality and forecasting.

1.1 Problem Background

Nowadays, there are several IoT-enabled systems are available which monitor air quality using gas and particulate matter sensors and then forecast air quality. We enlist some of the major challenges that come across these systems.

- Complexity in development of sensing nodes.
- Accuracy of forecasted data.
- Power efficiency is a great challenge for these systems as different sensors consume more power.
- The cost of different systems has increased due to high-end technology.

2 Related Work

We surveyed existing AQI monitoring systems. Table 1 represents a literature survey conducted on various air quality monitoring systems. V Gokul et al. [5] developed a low-cost air quality monitoring unit, but this sensing unit lacks energy efficiency as Wi-Fi consumes a lot of power. S. Dhingra et al. designed IoT-Mobair for AQI monitoring, but this solution was complex compared to other AQI monitoring systems. Similarly, various other AQI monitoring systems are listed in Table 1, along with pros and cons. As an outcome of this literature, we can say that a low-cost, less complex and accurate forecast system can be looked forward to designing.

The findings from Table 1 are as follows:

- No gas sensor exists that is 100% selective to a single gas [2].
- It is necessary to use instruments that employ analytical techniques to identify gases [3].
- Computational complexity is again one of the major issues in different air monitoring systems.
- Different sensors and processors use a lot of energy [4].
- More energy is spent on the measurement using an analog or digital sensor than on transmitting the acquired data.

Table 1. Literature survey of various air quality monitoring systems.

S no	Title	Summary	Pros	Cons
1	"An Environmental Air Pollution Monitoring System Based on the IEEE 1451 Standard for Low-Cost Requirements" [8] 2008	The main goal of this project was to build an environmental air pollution monitoring system (EAPMS) that could measure common air pollutant concentrations using a semiconductor sensor array and the IEEE 1451 standards, especially the IEEE 1451.2 standard	Low-cost implementation. Small size smart device. Real-time pollution detection	Chances of Error are High. Sensors are highly vulnerable to silicon-based chemicals
2	"Design and Evaluation of a Metropolitan Air Pollution Sensing System" [4] 2016	The architecture, prototype and evaluation of a low-cost participatory metropolitan air pollution sensing system have been described. Multiple portable sensing units are compared	Low-cost implementation. Real time pollution detection. Small size smart device	Metal oxide sensors are cheap but non-linear and unreliable. Do not have proper facilities and certification to store and handle toxic gases
3	"Urban Air Pollution Monitoring System with Forecasting Models" [5] 2016	Three machine learning (ML) algorithms are investigated to build accurate forecasting models	M5P outperforms the other algorithms. Multivariate enhance the prediction accuracy	Computation Power cost is more
4	"Implementation of a Wi-Fi based plug and sense device for dedicated air pollution monitoring using IoT" [7] 2016	Smart devices used to detect air pollutants using greater vancouver air quality index table and max operator accuracy, aggregation, consistency, and method reliability improve the air quality index calculation	Low-cost implementation. Small size smart device. Real time pollution detection	Wi-Fi cost is low but not less than Bluetooth technology

(*continued*)

Table 1. (*continued*)

S no	Title	Summary	Pros	Cons
5	"Internet of Things Mobile–Air Pollution Monitoring System (IoT-Mobair)" [2] 2019	The kit has been integrated with the mobile application IoT-Mobair that helps the user in predicting the pollution level of their entire route. Data logging can be used to predict AQI levels	Low-cost implementation. Small size smart device. Real time pollution detection	Complexity
6	"Lessons Learned from the Development of Wireless Environmental Sensors" [3] 2020	Up to 100 times More energy is spent on the measurement using an analog or digital sensor than on transmitting the acquired data	Eco-Friendly sensor. Energy harvesting hardware architecture	Sensors produce too much data; therefore, a large data center is needed. Higher development costs. The maximum discharge current of super capacitors
7	"Air Pollution Monitoring and Prediction using IOT and Machine Learning" [9] 2021	Linear Regression, Support Vector Regression, SARIMAX time series, GBDT ensemble and Stacking ensemble model are used to forecast the AQI values for the next five hours	SVM was giving the best accuracy for a smaller range of predictions	Longer range prediction is lacking

3   IoT Enabled Indoor Air Quality Monitoring Framework

In this paper, we proposed an IoT-based system for monitoring air quality in indoor environments. Our system aims to solve the challenges of complexity, accuracy, power efficiency, and cost-efficiency challenges. We focused on reducing the above challenges with the proposed system architecture shown in Fig. 1. This shows the major components involved in monitoring, reducing, and interfacing modules. AQI monitoring setup consists of various gas sensors (MQ 135, MQ 7, MQ 8, MQ6) along with optical dust sensor GP2Y1010AU0F [5]. We used Arduino Uno as an IoT programming board and 16 * 2 LCD to display AQI values. We proposed a solution to the forecasting accuracy problem using SARIMAX [6]. Time series model. Using an auto cut off algorithm, power efficiency can be reduced while programming the sensing node setup.

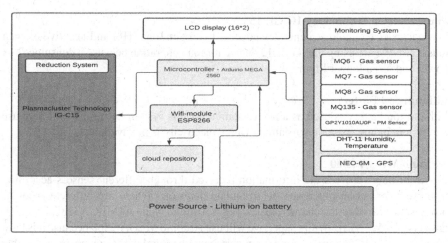

Fig. 1. Proposed system architecture for IoT based AQI monitoring

3.1 Major Components

Array of Sensor

Gas sensors are used to detect different pollutants which are available in the form of gases. GPS Sensor used to detect location anywhere on earth. Environment Sensor is used to detect conditions like temperature, humidity etc. Table 2 shows the different types of sensors used in our system.

Table 2. Hardware used in designing the setup

Type	Sensor	Use
Gas	MQ6	LPG, CO, and Smoke detection
Gas	MQ7	CO gas detection
Gas	MQ8	H_2 gas detection
Gas	MQ135	NH_3, NO and CO_2 gases detection
DHT	DHT 11	Detect humidity, temperature
PM	GP2Y1010AU0F	The concentration of particulate matter
GPS	NEO-6M	Sensing the geographical coordinates

Plasmacluster Technology-IG CL 15
Plasmacluster technology generates positive hydrogen ions (H^+) and negative oxygen ions (O_2^-). We are using the IG CL15 device for ion generation because it consumes less power and has a longer lifetime.

Lithium-Ion Battery
Lithium-ion battery is used as a power source for this system because it will help use our system in any indoor environment where electricity is not available.

Arduino MEGA 2560
Processor helps process the information received through different sensors according to those information processors to formulate decisions. There are different types of processors with different characteristics. E.g. - Arduino Nano ATMEGA328P, ESP8266 based NodeMCU-12E.In this paper, we will use Arduino MEGA 2560 because it is built upon an Atmega-2560 processor. In comparison with other boards, which are accessible easily, it comes with more memory storage and input/output pins.

Wi-Fi Module-ESP 8266
Wi-Fi is a communication method to connect processors and web servers with the help of the internet. We have used the ESP8266 Wi-Fi module because it is incorporated with TCP/IP protocol layers, which will allow microcontrollers to access any Wi-Fi network and connect with the internet.

User Application
It is a frontend web application where users can access the data of AQI, prediction of AQI, temperature, humidity, anything on a particular location.

3.2   AQI Forecasting Framework

Figure 2 represents our proposed AQI forecasting framework. The monitoring system is powered by a lithium-ion battery which supplies power to the sensing node, i.e. presented in Fig. 1. Data is collected from defined sensors and sent to the cloud repository using ESP 8266. Then data set is being created out of these sensor values, which act as training data set for further AQI forecasting. We have generated a prediction module based on the SARIMAX algorithm, which will help the system to forecast the AQI of a particular time. After forecasting, it is checked whether the current AQI is greater than the threshold. Then, with the help of the auto-cut-off algorithm request is sent to the plasma cluster to turn on and off.

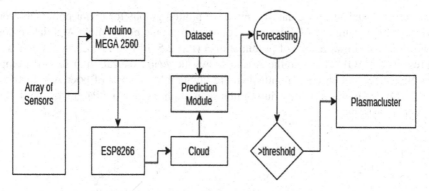

Fig. 2. Proposed AQI forecasting framework

4   Results and Discussion

AQI Monitoring and forecasting setup is shown in Fig. 3 From Fig. 3, we can see that the Step 1 is powering up the system with a lithium-ion battery. In the Step 2, different sensors, i.e., gas, PM, and temperature sensors, provide the sensed data to Arduino Mega 2560. The data received from sensors is sent to the thing speak cloud repository with the help of the ESP8266 Wi-Fi module, thus making it our Step 3. After data is sent to the cloud repository, in Step 4, our system applies the forecasting method over the received data sets and check whether the upcoming forecast value is less than the threshold or greater than the threshold. If the result is greater than the threshold, a request is sent to IG CL-15 to release ions. In Step 5, our LCD display shows the output of the current level of AQI.

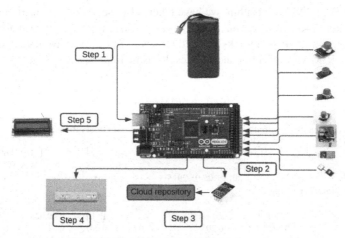

Fig. 3. AQI monitoring and forecasting setup

For forecasting we created an AQI dataset to as a training dataset for further forecasting of AQI values. We applied forecasting techniques such as ANN, RNN and SARIMAX time series models over the dataset. First, we plotted graphs for major pollutants from 2015 till 2020 using old AQI data sets collected from the government AQI data repository. Figure 4 shows results of pollutant data (PM 2.5, PM 2.5, NO2 and CO). Y-axis represents the AQI value, and the X-axis shows the years. We can infer from the graph that November to February months is more vulnerable in terms of poor AQI in India. The reason for poor air quality during this span might be due to "Parali" or fog during the winter seasons.

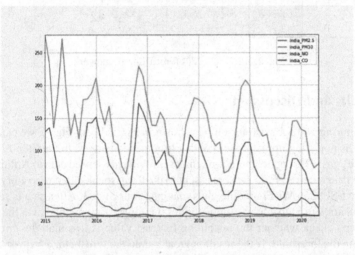

Fig. 4. Various pollutant values by year

For the SARIMAX algorithm, we have taken 41 observations for testing purposes. We have taken fewer observations because our purpose is to design an indoor system and to find if our algorithm works better than RNN, ANN. We have taken stationarity as 12 for SARIMAX algorithm and saw results as in Table 3.

Table 3. SARIMAX results

Dep. Variable:	India_AQI		No. Observations:	41	
Model:	SARIMAX(1,1,1)x(1, 0, 1, 12)		Log Likelihood	-204.793	
Date:	Mon, 06 Dec 2021		AIC	419.586	
Time:	10:11:52		BIC	428.030	
Sample:	01-01-2015 – 05-01-2018		HQIC	422.639	
Covariance Type: Opg					

	Coef	std err	Z	P>|z|	[0.025	0.975]
ar.L1	-0.6704	0.203	-3.309	0.001	-1.067	-0.273
ma.L1	0.9998	8.819	0.113	0.910	-16.285	18.285
ar.S.L12	0.9309	0.266	3.504	0.000	0.410	1.452
ma.S.L12	-0.6349	0.693	-0.916	0.360	-1.993	0.723
sigma2	1297.7696	1.13e+04	0.115	0.908	2.08e+04	2.33e+04

Then we focused on the data collected by the sensing node and applied ANN, RNN and SARIMAX forecasting models on the dataset. We plotted graphs of resultant AQI value as shown in Fig. 5. Y-axis represents the AQI value, and X-axis shows the months.

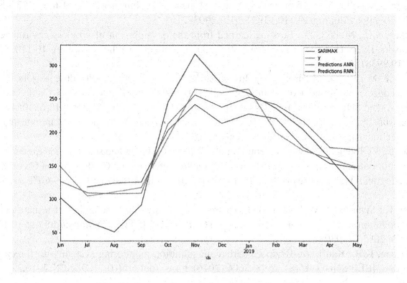

Fig. 5. Comparison of forecasting algorithms

Y represents the true values of our training data set. From Fig. 5, it can be inferred that if we are going with indoor AQI data prediction, then SARIMAX is giving the best forecasting results for AQI prediction.

5   Conclusion and Future Scope

We surveyed various IoT-based air quality monitoring setups. We designed a setup to measure AQI and monitored air quality. Further, we forecasted the value of AQI based on the data sets we created out of our sensing node. In conclusion, we can say that SARIMAX time series model is giving the most accurate results for forecasting AQI in longer durations, that is, AQI value by months. For future work, we can include Ozone sensor values and metrological components such as temperature and humidity in the framework to investigate the effect of these parameters on AQI. Also, we can look forward to implementing a plasma cluster mechanism to reduce air pollution.

References

1. Chen, X., Liu, X., Xu, P.: IOT-based air pollution monitoring and forecasting system. In: 2015 International Conference on Computer and Computational Sciences (ICCCS), pp. 257–260 (2015). https://doi.org/10.1109/ICCACS.2015.7361361

2. Kularatna, N., Sudantha, B.H.: An environmental air pollution monitoring system based on the IEEE 1451 standard for low-cost requirements. IEEE Sens. J. **8**(4), 415–422 (2008). https://doi.org/10.1109/JSEN.2008.917477

3. Dhingra, S., Madda, R.B., Gandomi, A.H., Patan, R., Daneshmand, M.: Internet of Things mobile-air pollution monitoring system (IoT-Mobair). IEEE Internet Things J. **6**(3), 5577–5584 (2019). https://doi.org/10.1109/JIOT.2019.2903821

4. Folea, S.C., Mois, G.D.: Lessons learned from the development of wireless environmental sensors. IEEE Trans. Instrum. Meas. **69**(6), 3470–3480 (2020). https://doi.org/10.1109/TIM.2019.2938137

5. Budde, M., Busse, M., Beigl, M.: Investigating the use of commodity dust sensors for the embedded measurement of particulate matter. In: 2012 Ninth International Conference on Networked Sensing (INSS), pp. 1-4 (2012). https://doi.org/10.1109/INSS.2012.6240545

6. Balasubramanian, S., Sneha, T., Vinushiya, B., Saraswathi, S.: Air pollution monitoring and prediction using IOT and machine learning (2021)

7. Gokul, V., Tadepalli, S.: Implementation of a Wi-Fi based plug and sense device for dedicated air pollution monitoring using IoT. In: 2016 Online International Conference on Green Engineering and Technologies (IC-GET), pp. 1–7 (2016). https://doi.org/10.1109/GET.2016.7916611

8. Hu, K., Sivaraman, V., Luxan, B.G., Rahman, A.: Design and evaluation of a metropolitan air pollution sensing system. IEEE Sens. J. **16**(5), 1448–1459 (2016). https://doi.org/10.1109/JSEN.2015.2499308

9. Shaban, K.B., Kadri, A., Rezk, E.: Urban air pollution monitoring system with forecasting models. IEEE Sens. J. **16**(8), 2598–2606 (2016). https://doi.org/10.1109/JSEN.2016.2514378

Journey of Database Migration from RDBMS to NoSQL Data Stores

Neha Bansal[(⊠)], Kanika Soni[(⊠)], and Shelly Sachdeva

Department of Computer Science, National Institute of Technology, Delhi,
New Delhi 110040, India
{nehagoel,kanikasoni,shellysachdeva}@nitdelhi.ac.in

Abstract. Migration is a complex process involving many challenges while migrating from an existing system to a new one. Database migration involves schema transformation, migration of data, complex query support, and indexing. This paper presents a) Journey on existing migration techniques from RDBMS (SQL) to NoSQL databases. Schema migration and Data migration are two main aspects while migrating from relational to NoSQL database. b) The various existing techniques related to schema migration and data migration. The techniques presented for schema migration include graph-based algorithms, denormalization techniques such as column level denormalization and table level denormalization, link list-based, and ETL tools. The paper also addresses different business methods for data migration, commonly known as basic tools and ETL tools. c) The authors highlight some open challenges like need and cost of denormalization and selection of columns for denormalization. The assessment of various techniques (based on space or time costs) is presented. The key challenge is to pick a particular datastore on which to apply a specific technique. d) The paper also describes current market-driven migration tools based on each particular data store. e) It also throws light on different organizations that successfully migrated to a particular NoSQL data store. Thus, the purpose of this study is to contribute to the state-of-the-art in the field of database migration and to serve as a foundation for selecting and developing RDBMS-to-NoSQL data migration techniques or tools.

Keywords: Database migration · NoSQL · Relational database · Schema denormalization · Data migration

1 Introduction

The development of databases has experienced a long database evolution journey, starting from hierarchical, network, relational, object-oriented, NoSQL to NewSQL [7]. The growth of the database technologies results from the changes in the users and the development needs, increasing hardware capabilities, and the development of other IT technologies. Before the big data [38] need, relational databases were the most popular and had been the market leader for over thirty years. With the growth of internet users, web applications, e-commerce, social media network, and hardware capabilities, we have an ever-increasing amount of data. This data comes in different formats, stored and

© Springer Nature Switzerland AG 2022
S. Sachdeva et al. (Eds.): BDA 2021, LNCS 13167, pp. 159–177, 2022.
https://doi.org/10.1007/978-3-030-96600-3_12

analyzed according to development needs, and can be used to make decision support systems, predictive analysis, and many other advantages to the IT industry. Many organizations realize that the database's trivial system is insufficient to store and manage this massive amount of data. This problem pushes relational databases beyond their limits, like more budget strain, frequent schema change, and scalability issues. These limitations compel many organizations to migrate from a relational database to a new type of database called NoSQL, which can handle these limitations. Using a NoSQL database over a relational database includes high scalability, simple design, and high availability with more precise control.

1.1 NoSQL Background

NoSQL means 'Not Only SQL' [2, 3] is a highly optimized key-value store to achieve performance benefits in latency and throughput. It provides many essential features [4]: distributed nature, horizontal scaling, data partitioning, data replication, open-source, weaker consistency model, and support for simple API (Application Interface). These properties make NoSQL more accessible and different than conventional database technology (like a relational database). NoSQL database is further classified [5] into four types, i.e., document, key-value, column-based, and graph stores. These four types have similar logical organization: a key (system generated or user derived) followed by a value but a different physical organization like data modeling, data architecture, query languages, API. Table 1 shows the four different types of NoSQL [36] database along with the basic description, their use case, popular open-source stores, and their current industrial use.

Table 1 shows the NoSQL databases along with the basic description of each type, their use case, popular data stores, and their current industrial use. According to Table 1, key-value stores are the purest form of NoSQL stores, which takes a key and stores the value corresponding to the key, as shown in Fig. 1b. These stores are known as highly performant, efficient, and scalable. Facebook, Instagram, Pinterest are the companies that use key-value stores. Document stores work by storing the information in JSON documents, as shown in Fig. 1c. It also gives an API and query language to access and process the data. The primary use cases are web-based analytics, real-time analytics, and e-commerce applications. The most popular known stores are MongoDB, CouchDB, and DynamoDB, which big companies like Twitter, eBay, and MetLife use. The use of Column data stores is to store data, as seen in Fig. 1d, in column families. They are primarily used to process large numerical data sets batch-oriented and suitable for handling large quantities of unstructured or non-volatile data. The use cases are log aggregation and blogging platforms. This category's known data stores are Apache Cassandra, HBase, and ScyllaDB, which are used by many well-known companies like Accenture, Twitter, Facebook, Netflix, and many more. Graph data-stores store data in the form of a graph where nodes represent data, and the edges represent the relationship between them, as shown in Fig. 1e. It provides excellent read efficiency where data is highly interconnected. The primary use areas of such databases are a need to access highly inter-related data and visit many relationship levels like social networks and geospatial data. The common data stores for the graph are Neo4j, AllegroGraph. Twitter, Target, Addidas, eBay, and even more use such data stores.

Table 1. Various categories of NoSQL stores and their Industry use

S. no.	Databases	Description	Suitable cases	Popular database name	Industry/Company
1.	Key-value stores [18]	• Stores data in simple key-value pairs	• Caching, • Session Management • Message queuing	• Redis [19] • Memcached [20] • Riak [21]	• Facebook • Instagram • Twitter • Pinterest
2.	Document Stores [18]	Stores data in the form of documents containing field and their values, e.g., JSON format	• Web-based analytics, • Real-time analytics, • e-commerce applications	• MongoDB [26] • DynamoDB [27] • CouchDB [25]	• Twitter • eBay • MetLife • Shutterfly
3.	Column family stores [18]	• Stores data in column families	• Log • Aggregation, • Blogging platforms	• Cassandra [22] • HBase [23] • ScyllaDB [24]	• Accenture • Facebook • Netflix • Pay pal
4.	Graph databases [18]	• Stores data in the form of a graph where nodes represent data and the edges represent their relationship	• Social networks, • Geospatial data	• Neo4j [28] • AllegroGraph [29]	• Facebook • Walmart • Adidas • Telenor • eBay

Table 2 presents a comparison between RDBMS for various types of NoSQL databases. The purpose of the analogy is that relational databases have been well known for a very long time while NoSQL databases are new in the market. The translating terminology can be helpful for novice users to understand these new data stores. In Fig. 1, the authors take an example according to the above analogy. A relational table stores the details of various office locations situated in different countries of the world, as seen in Fig. 1a. The respective figures demonstrate how four types of the NoSQL database will map the same information in the relational table. Figure 1b gives the representation in key-value stores. Here we have taken the country name as a key and store the value corresponding to it as values.

Figure 1c shows how document stores work by saving the same information in collections and documents. Here tables in RDBMS are stored as documents. For example, given two documents are created named document 1 and document 2. Both of the documents are inside a collection. Figure 1c presents the relational database representation

Table 2. The analogy between RDBMS and NoSQL data stores

Relational model	Key-value store	Document store	Column family store	Graph store
Database	Database	Database	Keyspace	Database
Table	-	Collection	Column family	Label on nodes
Row	Key-value pair	Document	Row	Node
Column	-	Key	Column	Node properties

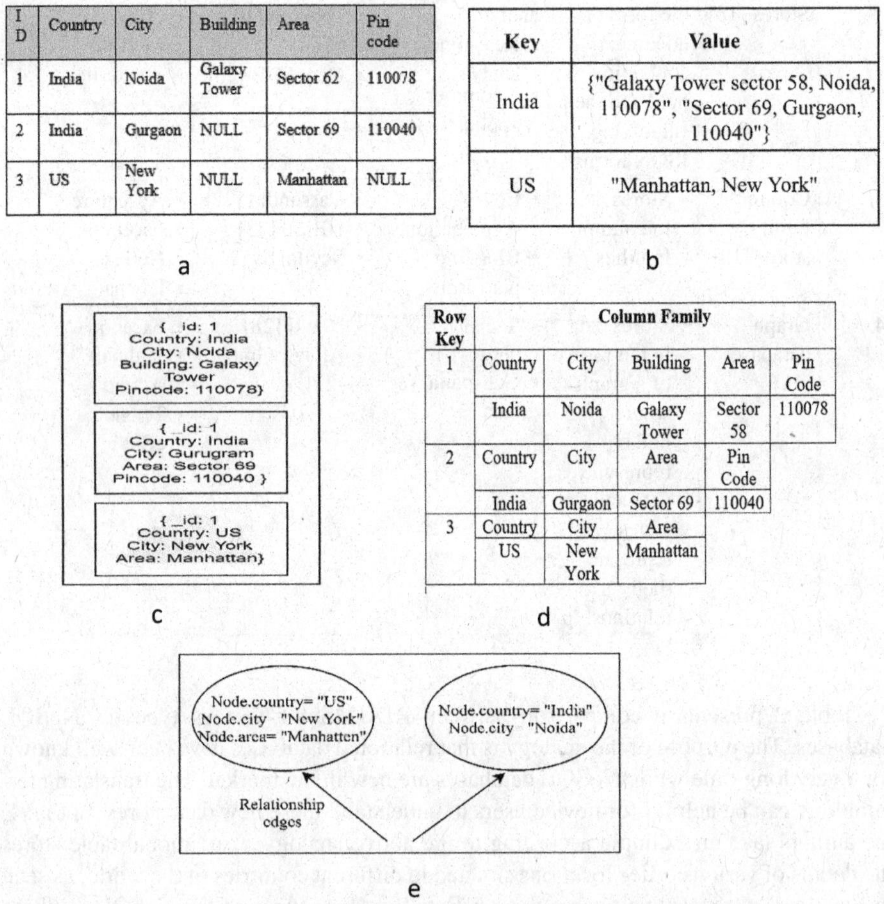
ID
Country
City
Building
Area
Pin code
1 India Noida Galaxy Tower Sector 62 110078
2 India Gurgaon NULL Sector 69 110040
3 US New York NULL Manhattan NULL
a
Key
Value
India
{"Galaxy Tower sector 58, Noida, 110078", "Sector 69, Gurgaon, 110040"}
US
"Manhattan, New York"
b
{ _id: 1
Country: India
City: Noida
Building: Galaxy Tower
Pincode: 110078}
{ _id: 1
Country: India
City: Gurugram
Area: Sector 69
Pincode: 110040 }
{ _id: 1
Country: US
City: New York
Area: Manhattan}
c
Row Key
Column Family
1 Country City Building Area Pin Code
India Noida Galaxy Tower Sector 58 110078
2 Country City Area Pin Code
India Gurgaon Sector 69 110040
3 Country City Area
US New York Manhattan
d
Node.country= "US"
Nodc.city "New York"
Nodc.arca= "Manhatten"
Node.country= "India"
Nodc.city "Noida"
Relationship edges
e

Fig. 1. Data stored in RDBMS and corresponding NoSQL formats; a) Relational Table; b) Key-Value Store; c) Document Store; d) Column Family Store; e) Graph Store

in column stores. It takes a row key and stores the related information in the column corresponding to the key. The country name is used as a row key for the given case, and the column stores the relevant information. Likewise, Fig. 1d demonstrates how graph stores are operated in graph form to keep the details. For each country name, the graph database generates two nodes for the specified example. Figure 1 illustrates how the physical architecture of a NoSQL database differs from that of a relational database. While migrating the data, it is necessary to change the physical architecture, which is a tedious task to perform manually.

Database migration moves the business logic, schema, physical data, and database dependencies from a current system to a different system. It may be because of various reasons such as cost, capabilities, functionalities, requirements. For example, a company can consider saving money by moving onto cloud systems. Another example may be moving the existing electronic health records data (usually persisted in RDBMS). It can find that some particular platform is more accurate to fulfill its users or business needs, or the present system is simply outdated. Database migration is a complicated process and involves various steps named schema migration, data migration, and query translation and optimization (detailed in the next section) to ensure no loss of information and that data security is maintained.

The article presents the following parts. Section 2 explains the issues, the need for migration. Section 3 points out the different techniques and migration models along with the pros and cons of all the models. Section 4 presents the existing migration tool in the market. Section 5 gives the related work. Ultimately, Sect. 6 discusses numerous migration study problems, and the last section concludes the article.

2   A Journey of Migration from RDBMS (SQL) to NoSQL

The process of migration is not just declining the existing platform and moving onto the next one but also involves data migration, schema changes, solutions for querying, indexing along the functionalities of different NoSQL systems are to be kept in mind. Figure 2 shows the database migration process from relational to NoSQL database. RDBMS consists of three essential parts, i.e., schema, SQL queries, and workload. While migration from RDBMS, these parts need to be considered and require proper methodologies to move into a new system. So, database migration consists of three phases named as a) schema migration, b) data migration, and c) schema translation and optimization. However, this paper only focuses on two main phases, i.e., schema migration and data migration.

Schema Migration: The transformation of the existing relational tables' schema into the corresponding NoSQL database's data model is called schema migration. The schema migration phase consists of the migration of tables, columns, primary keys, foreign keys, indexes, views, etc., from relational systems to the NoSQL system. The problems while schema migration is a) Relational database stores the data in a normalized way, whereas NoSQL stores information in a denormalized way. b) The relational database follows a rigid schema, whereas NoSQL provides flexibility in the schema. c) Relational database support joins, whereas NoSQL avoids joining. So, schema migration is a complicated phase because no data should be lost while migration.

Data Migration: It is the process of migrating the relational tables' data into the underlying NoSQL database. The data migration involves the transform and load steps in which data gets converted from the relational database according to the NoSQL database's underlying data model. Then the data is finally loaded into the NoSQL database. Data Migration phase is also crucial during migration because this phase may introduce some errors like redundant or some unknown data in the destination system.

Query Translation and Optimization: NoSQL does not support complex SQL queries. These complex queries need to be translated into some simplified queries using query unnesting techniques [39, 40] and temporary tables. After the complex query is simplified, the uncomplicated queries need to be optimized to execute in less time.

2.1 Need for Migration

Relational databases are adopted widely for data-intensive domains and transactions. With the rise of internet and web technologies, data is growing faster, and handling this large and complex data with RDBMS has become a challenging task. Big companies like Google, Facebook, Microsoft, and many more want to use this vast data to benefit business and money. New technologies like NoSQL can quickly achieve the benefits. The following points elaborate the need for migration from SQL to NoSQL [1, 2]:

a) **Handling Big Data:** With the big data wave in the last decade, development and data processing needs have changed. The hardware capacity is growing day by day, for which we need a storage methodology that should enhance the proper utilization of big data.

b) **The emergence of Workload:** Two kinds of workload emerged:

- **Bullet queries** (need for availability) in interactive data serving domains like gaming and the social network that operates (generally reads operations) on a tiny part of the database for which the whole transaction process is overhead, as they need fast availability.

- **Decision support systems** (Need for analytical support) need analytical resources on a massive amount of data. NoSQL systems like document-based stores and Hadoop-like systems for analytics have proved better in performance for such workloads.

- **Need for schema flexibility:** a new type of data emerging which can be semi-structured or unstructured, rather than just structured, maybe evolving constantly or sometimes is sparse; hence the need for the flexibility of schema has also become important.

- **Need for scalability:** Application nowadays is not limited to a finite audience, data needed to be made available at high speed to a large number of users, shifting towards clouds, or having multiple servers instead of large monolithic servers has become trending; therefore, the databases today need to focus on providing good scalability.

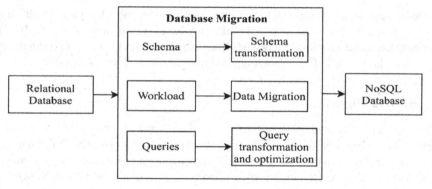

Fig. 2. The database migration process from RDBMS (SQL) to NoSQL data stores.

2.2 Issues

Many aspects need to be considered while Migrating from the relational database to the NoSQL data store [8]. **Different data models**: Relational and NoSQL technologies are different from each other. Relational databases store data in a completely structured way, whereas NoSQL is entirely unstructured. These two technologies have different data models. **Selection of Modeling techniques:** NoSQL has four different data models and to choose one among them is very challenging. When deciding which NoSQL model to employ, it's a good idea to determine how you plan to use the data and then pick the NoSQL model most suited for that job. **No support for 'join' operation**: NoSQL databases do not support the 'join' process, although some databases like MongoDB have started offering references in general, based upon "CRUD" (create, read, update and delete). Hence, while migrating, we need to ensure that the read operations are efficient after the migration. There should not be an overhead of reading from multiple places. **No query interface**: NoSQL databases do not provide an easy-to-use interface for querying; the queries in the applications need to access the data. Hence while migrating, we need to keep in mind the query translation. **There is no support for efficient query processing and optimization:** Codes for efficient query processing and optimization have to be written by the developers. **Data Storage Approach**: The first challenge that needs to be faced is that RDBMS supports a decentralized approach, whereas NoSQL data stores follow the data distribution property.

As stated previously, the overall database migration process is a highly complex task that entails two phases of migration and numerous migration challenges. As a result, there is a need to develop tools that can complete the process quickly and accurately. Migration mechanisms are beneficial when selecting or creating the tool.

3 Migration Techniques

To do the database migration, there are many existing migration techniques in the software industry that can be helpful to make the migration process easy and error-free. Figure 3 presents the current schema and data migration techniques. As shown in Fig. 3, schema migration consists of four techniques named as a) Graph transforming algorithm,

b) Denormalization techniques which further subdivides as Table level denormalization and Column level denormalization, c) Linked list-based schema migration, and d) ETL (Extract, Load and Transform) tools, whereas data migration consists of two significant techniques known as a) Basic data migration tools and b) ETL based tools.

3.1 Schema Migration

The schema of relational databases maps onto the schema of a target database model. Relational databases use the schema normalization technique, whereas NoSQL databases use the schema denormalization. Due to this problem, mapping is not an easy and straight task. Many methods exist to transform schema from a relational database to NoSQL. In this paper, as shown in Fig. 3, the primary four methods are discussed.

3.1.1 Graph Transforming Algorithm

Graph transforming algorithm [3, 9] is a schema conversion technique that uses the graph data structure to convert schema from SQL to NoSQL database. SQL stores data in tables and the related dependencies are represented by the foreign keys, whereas NoSQL stores data by datasets independent of each other. The method transforms the schema into a graph model where nodes represent tables and edges represent the relationship between tables. The extension rules extend the tables into one another and do the table nesting to preserve the relationships used.

3.1.2 Denormalization

One of the simplest ways for schema transformation is to map the schema in one to one manner, known as the normalization method [5]. It often results in poor query performance for NoSQL due to the absence of join operations where reading from different tables becomes an overhead. Hence, denormalization is used for schema transformation. Denormalization is combining one or more tables or attributes or so to improve the read efficiency and reduce the response time. But denormalization might increase the redundancy in data and inefficiency in update operations. Hence, the algorithms designed around denormalization need to consider a trade-off point between reading from different places and redundancy, which leads to inefficient updates. There are two types of denormalization stated as Table Level Denormalization and Column Level Denormalization.

a. **Table Level Denormalization**
   A simple method to avoid join operations by merging all tables into a single table along with primary-foreign key relationships is known as Table Level Denormalization [5]. This method uses the graph data structure known as a schema graph to represent a primary-foreign key relationship between various tables. The schema graph is converted into a schema tree. Different tables of RDBMS are then combined into one aggregate according to the schema tree so that data is always read from one location.

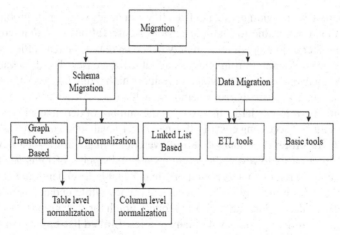

Fig. 3. Existing migration techniques

b. **Column Level Denormalization with Atomic Aggregates**

The column-level denormalization approach [5] repeats only a few columns instead of entire tables. This method maintains all of the original tables, duplicates only a few columns. The biggest challenge here is to select only different columns to repeat. This method duplicates only columns that appear in non-primary-foreign-key-join predicates and makes query graphs for each query. This method uses atomic aggregates to support transaction-like behavior in NoSQL applications. Those tables that get modified together in a transaction are combined based on their relationship, which could be one-to-one, many-to-one, or one-to-many.

3.1.3   Linked List Based

The linked list-based method [15] follows DDI principles: Denormalization, Duplication, and Intelligent keys. The SQL tables are re-aggregated into a big NoSQL table even if some data is redundant in the NoSQL table. Then, an intelligent key called row key is selected for each row to identify it uniquely. This method works on the NoSQL column family store.

3.1.4   ETL Tools

Extract Transform Load tools [16] extract the schema from RDBMS tables, transform it to equivalent NoSQL schema, and finally, loads the data to the target NoSQL schema according to the new schema. The tool is basically used in the migration of data warehouses.

Table 3 gives the comparison among these primary four techniques used for schema migration. The comparison is made on three parameters named NoSQL Category, which states which type of NoSQL the technique is applicable, Advantages of each technique, and challenges associated with each technique, which are still an open problem and give future research directions. Based on the comparison, it is stated that **Graph Transforming Algorithm** is applicable to document stores with the advantage of improving computer output by providing excellent reading efficiency. The open problem is that it takes up much space due to multiple nesting steps. **Table-Level denormalization** reduce the number of join operations, mainly hence increases the read efficiency. It also adds the overhead of reading from many small tables decreases. The disadvantage is a) It stores the data in one large table, which introduces high redundancy resulting in heavy update operations b) Needs extract operations, which are always costly. c) It doesn't support atomicity d) Needs large disk sizes as the number of dependencies increases e) High processing and accessing cost of information retrieval due to massive I/O memory operations. **Column-level denormalization** overcomes many of the table-level denormalization limitations. However, shortcomings are: a) Number of joins needed is less than the normalized relational table but more than the table-level denormalization b) As the columns duplicate, this might lead to inconsistency while update, as it has two to updated at all the places, duplicated c) A new query graph for each graph hence increases complexity for real-time systems with many queries. **Linked list-based** only works for column family stores. It works well for analytical queries and offline data processing, but this method performs poorly for real-time applications. **ETL tools** are quite famous for migration because they support schema and data migration operations as a single step. Still, it sometimes results in update anomalies, which can hinder a system's performance and do not provide the facilities of customization during the conversion process.

3.2 Data Migration

Data migration is the process of migrating the data of a system or application from one database server to another. Data migration is performed after the schema migration. Tools are used to do the data migration, which falls into either of two categories named Basic tools and ETL tools. These tools are basically developed by various large organizations or by academic researchers.

3.2.1 ETL Tools

ETL [11] is used worldwide for the development of data warehouses. It involves three tasks named as the extract, transform and load. There are numerous ETL tools [13] available for performing ETL tasks. The tools [11] can be classified as data integration tools, code-based tools, cloud-based tools, and real-time tools, among others. There are dedicated NoSQL ETL tools such as MongoSyphon, Transporter, Krawler, Panoply, Stitch, Talend Open Studio, and Pentaho that are designed to work with NoSQL data stores.

3.2.2 Basic Tools

These are the simple tools use to migrate data from one platform to another. The purpose of using these tools is to transfer bulk amounts of data in little time. All the space taken by these tools is significantly less, and they provide excellent read efficiencies. Amazon, Google, and Microsoft offer the commercial and open-source types of tools for different categories of NoSQL. The various available tools are Apache drill, Informatica, and many more.

Table 4 gives the advantages and open challenges of data migration techniques. ETL tools, as mentioned above. Some basic tools provide advantages like excellent read efficiency, bulk data transfer, and less occupation of memory space but suffer from challenges like most tools offer no solution for importing primary-foreign key relationships between tables of the relational database to NoSQL. But can sometimes lead to error-prone, inconsistent data, and high downtime.

4 Existing Migration Tools in the Market

NoSQL technologies are new in the market in comparison to relational databases and still evolving in nature. NoSQL technologies sacrifice consistency over many advanced features like flexibility, scalability, and high availability to handle Big data. Many big companies migrate from RDBMS to NoSQL data stores to take the OLAP data, not involving too many transactions. Top NoSQL data management systems like MongoDB, Neo4j, Cassandra, and Redis have developed their migration tools to make the migration process easy.

Table 5 shows the existing open-source tools for enterprise-scale data export and import, also known as migration tools for the NoSQL database. Also, a particular tool supports only one specific datastore of a specific type. As given in Table 5, tools like the Redisco python library [13] are designed to migrate from RDBMS to Riak. There are numerous ETL [10] tools developed by different vendors. The ETL tool used for MongoDB is MongoSyphon [10], for Cassandra is Kettle [11], whereas for CouchDB is Talend [30]. For DynamoDB, the existing migration tool is AWS Database Migration Service (AWS DMS) [31] can migrate data to and from Open-Source databases, which involves not too complex data migration. Use EMR, Amazon Kinesis, and Lambda with custom scripts [31] method when more flexibility and complex conversion processes and flexibility are required. Load CSV [12] is the simplest method for migration from RDBMS to Neo4j. APOC [12] is developed as a library to give common procedures and functions to the users. ETL tools [12] provide user interface (UI) tools for migration using JDBC connections. The kettle is an enterprise solution for data import and export. Other ETL tools provide a GUI-based interface by different vendors for migration. Programmatic via drivers [12] can retrieve data from a relational database (or other tabular structure) and use the bolt protocol to write it to Neo4j through one of the drivers with your programming language of choice.

Table 6 illustrates that many enterprises belonging to different platforms have successfully migrated from RDBMS to different NoSQL databases using the various tools given in Table 5. The migrated companies list is very long, but this paper covers some of the popular companies that have successfully migrated from relational databases to some popular NoSQL data stores. The information is taken from the official websites of popular data stores like MongoDB, Cassandra, Riak, and Neo4j.

5 Related Work

Because of NoSQL's success, a great deal of work has been done on integrating NoSQL into current systems or switching from SQL to NoSQL. For the adaptation of the application to NoSQL, the solutions [37] have been categorized into

- Redevelopment; which is about remodeling the app from scratch.
- Wrapping; which focuses on providing a new interface to the current software components to exploit modern technology's advantages.
- Migration aims to transfer the current application into a new setting while maintaining the original functionalities.

The migration process [17] consists of three divisions: i) Online migration with continuous synchronization; in which the application is always working and remain connected to the database in parallel with using resource database for the process of migration, ii) Offline migration through tool; in which the application is completely down, and the tools use all the computation power and resource database for migration, iii) Manual migration through scripting; in which the form is entirely down, and the developer uses all the computation power and resource database to execute the script. There has also been much work done to provide interfaces to the current Relational database to use the NoSQL advantages. For example, at the top of the Relational Databases is a JSON layer [14]. Since JSON is the most common format for Web 2.0 applications, developers can use JSON to access and recall data. Further work is needed to ensure that NoSQL does not have a proper query language and processing or optimization of queries, so either developer has to write the codes themselves after migration or some configuration or middleware has to be used purpose. The extensible middleware [4] helps to translate SQL queries to the connected NoSQL systems. The migration between these systems is thereby greatly simplified. Of course, only a subset of SQL operations can be offered due to the NoSQL stores' limited functionality.

Karnitis et al. [40] use a two-level approach using a physical level model of relational schema and a logical level model for the schema transformation. Still, the templates or business concepts are decided by the users only. Jia et al. [39] consider the frequently joined relations and modified to determine whether to increase the redundancy by embedding documents or giving references and making the updates efficient. Yuepeng Wanget et al. [41] has designed a Datalog tool to migrate the data between the Relational database to NoSQL graph or document stores. They have tested the tool on different benchmarks and successfully migrated the database from one platform to another. Toufik Fouad et al. [44] have given a methodology to migrate schema and data from the Object-relational database to column stores (Cassandra) with the help of some schema transformation rules. However, the authors only give the theoretical concept. Yelda Unalet et al. [43] has provided a methodology for both schema and data migration from the Relational database to a graph database. They have given rules to migrate from tables into graphs. But the only theoretical concept is given. Diego Serrano et al. [42] has presented some set of rules to migrate from Relational Database to column stores (HBase). It follows a four-step data schema transformation process for schema migration from the ER diagram model onto HBase's multidimensional maps. It follows the aware query method for mapping SQL queries to the HBase API call sequence.

Table 3. Advantages and challenges of schema migration techniques

S. no.	Technique	NoSQL category	Advantages	Challenges with technique
1.	Graph transforming algorithm	Document store	• It provides excellent read efficiency	• The spatial redundancy is too much
2.	Table-level denormalization	Document store	• Less the number of join operations • Reading overhead	• Costs heavy update operations • Needs extract and unwind operations • Don't support atomicity • Needs large disk sizes • High processing and accessing cost
	Column-level denormalization with atomic aggregates	All	• Low cost of data accesses due to support of unwind operation • Improves the read efficiency • Provides atomicity • Medium usage of disk and accessing cost	• More number of join operations than the table-level denormalization • Data inconsistency due to column duplication • Query graph generation increases complexity for real-time systems with many queries
3.	Linked list based	Column family	• Reduces read overhead due to denormalization • Suitable for Analytical data	• Not suitable for real-time data transactions where time is the cost
4.	ETL tools	All	• A single tool that does schema and the data migration on a single go • Used by all popular NoSQL DBMS	• ETL tools based on denormalization suffer from data redundancies and lead to inefficient update operations • Do not provide customization facility

Table 4. Advantages and open challenges of data migration techniques

S. no.	Technique	NoSQL category	Advantages	Challenges
1.	Basic	All	• Tools are applicable to many NoSQL types • Easy to work • Customizable	• Error prone data • Inconsistent data • High downtime
2.	ETL tools	All	• Used for data warehouses	• ETL tools suffer from data redundancies and take more storage space

Table 5. Existing migration tools of some accessible NoSQL Databases

Category	NoSQL DBMS	Migration tools
Key-value store	Redis	Redisco python library
Document store	MongoDB	MongoSyphon (Extract Transform Load Tool)
	DynamoDB	AWS DMS Use EMR, Amazon Kinesis, and Lambda with custom scripts
	CouchDB	Talend (Extract Transform Load Tool)
Column store	Cassandra, HBase	Sqoop ETL tools such as Pentaho's Kettle
Graph store	Neo4j	LOAD CSV APOC Internally built ETL Tool Kettle Other ETL tools Programmatic via drivers

Table 6. Use cases of migrating to different NoSQL data stores.

S. no.	Category	Company name	Migration platform	Reason
1.	Key-value store [32]	Bump	MySQL to Riak	To improve opertional quality like scalability, capacity, and performance
		Dell	MySQL to Riak	To provide cross-data center redundancy, high write availability, and fault tolerance
		Alert Logic	MySQL to Riak	To perform real-time analytics, detect anomalies, ensure compliance, and proactively respond to threats
		The weather company	MySQL to Riak	To support their massive significant data needs

(*continued*)

Table 6. (*continued*)

S. no.	Category	Company name	Migration platform	Reason
2.	Document store [33]	Shutterfly	Oracle to MongoDB	For high performance, scalability and fast response time to market users
		Berclays	MySQL to MongoDB	For agility, performance, cost-effectiveness
		Cisco	RDBMS to MongoDB	For high performance, fast response
		Sega	RDBMS to MongoDB	High performance, scalability, and fast response
3.	Column store [35]	Coursera	MySQL to Cassandra	Due to unexpected downtime and single point of failure
		Mahalo	MySQL to Cassandra	For cost reduction, scalability, and high performance
		Scoop.it	MySQL to Cassandra	To handle extensive data, high performance
		Zoosk	MySQL to Cassandra	For their high volume of writes of time series data
4.	Graph store [34]	Digitate	PostgreSQL to Neo4j	Handling of the massive interconnected amount of data
		Fortune 50 retailer	DB2 to Neo4j	To handle complex searches among masses of connected data

6 Discussions

Migration is a complex process and needs many intermediate phases before successful migration. This paper discussed many of the existing migration techniques (such as Graph transforming, TLD, CLDA), their advantages, and the challenges faced. This section discusses various research problems while improving existing methods.

Research Problem (RP) 1-Is it essential to renormalize all the related tables into a single table?

Denormalizing all the related tables (foreign-primary key relationships) into a single table may create massive redundancy with high space complexity. Furthermore, it is challenging to denormalize enormous data into a single table in a reasonable period. Also, update queries can become complicated. Hence, it is not advisable to renormalize everything. Then, CLDA solves the problems faced by TLD.

RP 2-CLDA makes a transaction query graph for each query. Then, how will it face issues in a real-time system where queries vary user to user?

In a real-time system, there can be a variety of queries demanded by the user. Thus, creating a query graph for each query every time is not feasible by CLDA. It may take more time to handle all the queries.

RP 3-Can the CLDA method be applied to other NoSQL data stores?

CLDA [6] applies to Apache H-Base, a NoSQL column store on top of Apache Hadoop. CLDA worked for the NoSQL document store. But, Kim, Ho-Jun, etc. [8] claims that it applies to other NoSQL data stores.

RP 4-Graph transforming algorithm works for document stores. Can it be applied to other data stores, and can the space complexity be reduced?

As the paper gives, the graph transforming algorithm is applied only on document stores but with changes according to a particular data model. It works for other data stores. Reduce the space complexity is still an open problem for the graph transforming algorithm.

RP 5-Can the Linked list-based migration method be applied to other columnar data stores like Cassandra?

Given the authors, the method is applied only to HBase, and there is no proof to use it on other column data stores. Whether it can be applied or not is still an open problem for research.

RP 6-Can data migration provide customization to the users to choose schema and preserve the data dependency

The data migration technique can be independent of the schema transformation. Authors [12] discuss a metamorphose framework in which they made the data migration more efficient by allowing UDFs and customization by the users for the choice of the schema. For other datastore tools, the problem is still open.

7 Conclusions

With the advent of big data, conventional databases like RDBMS can't accommodate data volume and variety, and the modern database era calls for architecture solutions to be revamped. So, to fulfill the new data need, NoSQL databases come into the picture. They provide many advanced features like flexibility, scalability, better performance, and distributed data. The paper outlines the need for migration, migration problems, and the challenges confronting RDBMS migration to NoSQL stores. NoSQL has four distinct subcategories having different physical architecture. Giving a standard migration solution that can work equally well for all NoSQL categories is very challenging. Database migration involves two main phases, i.e., schema migration and data migration phase. Both phases are very critical to the efficient migration of the database. There are many existing techniques for each phase in the market. The paper discusses the graph-based algorithm, denormalization, linked list-based, and ETL tools for schema migration. Basic tools and ETL tools for discussion related to data migration. Also, each technique has some open problems and challenges which are still unanswered and still need further research to overcome those challenges. Due to various limitations of RDBMS, many big organizations have successfully migrated from traditional systems to NoSQL databases with the help of existing migration tools to improve the overall performance and fulfill the evolving data need. However, all of the existing tools are not open-source and require expert knowledge to operate. In the future, we plan further to analyze the open challenges of the strategies mentioned and explore options for overcoming these challenges within a systematic framework for database migration.

References

1. Floratou, A., Teletia, N., DeWitt, D.J., Patel, J.M., Zhang, D.: Can the elephants handle the NoSQL onslaught? Proc. VLDB Endow. **5**(12) (2012). https://doi.org/10.14778/2367502.2367511

2. Özcan, F., et al.: Are we experiencing a big data bubble? In: Proceedings of the 2014 ACM SIGMOD International Conference on Management of Data, pp. 1407–1408 (2014). https://doi.org/10.1145/2588555.2618215

3. Zhao, G., Lin, Q., Li, L., Li, Z.: Schema conversion model of SQL database to NoSQL. In: Ninth International Conference on P2P, Parallel, Grid, Cloud and Internet Computing, IEEE, pp. 355–362 (2014). https://doi.org/10.1109/3PGCIC.2014.137

4. Rith, J., Lehmayr, P.S., Meyer-Wegener, K.: Speaking in tongues: SQL access to NoSQL systems. In: Proceedings of the 29th Annual ACM Symposium on Applied Computing, ACM (2014). https://doi.org/10.1145/2554850.2555099

5. Yoo, J., Lee, K., Jeon, Y.-H.: Migration from RDBMS to NoSQL using column-level denormalization and atomic aggregates, J. Inf. Sci. Eng. **34**(1) (2018). https://doi.org/10.1007/978-981-10-6520-0_3

6. Kim, H.-J., Ko, E.-J., Jeon, Y.-H., Lee, K.-H.: Techniques and guidelines for effective migration from RDBMS to NoSQL. J. Supercomput. **76**(10), 7936–7950 (2018). https://doi.org/10.1007/s11227-018-2361-2

7. Davoudian, A., Chen, L., Liu, M.: A survey on NoSQL stores, ACM Comput. Surv. **51**(2), Article 40, 43 p (2018). https://doi.org/10.1145/3158661

8. Kim, H.-J., Ko, E.-J., Jeon, Y.-H., Lee, K.-H.: Migration from RDBMS to column-oriented NoSQL: lessons learned and open problems. In: Lee, W., Choi, W., Jung, S., Song, M. (eds.) Proceedings of the 7th International Conference on Emerging Databases. LNEE, vol. 461, pp. 25–33. Springer, Singapore (2018). https://doi.org/10.1007/978-981-10-6520-0_3

9. Sun, W., Fokoue, A., Srinivas, K., Kementsietsidis, A., Hu, G., Xie, G.: Sqlgraph: an efficient relational-based property graph store, In: Proceedings of the 2015 ACM SIGMOD International Conference on Management of Data, ACM, May 2015, pp. 1887–1901. https://doi.org/10.1145/2723372.2723732

10. RDBMS to MongoDB migration guide, MongoDB white paper (2018). https://www.mongodb.com/collateral/rdbms-mongodb-migration-guide. Accessed 10 May 2020

11. Kjellman, M.: MySQL to cassandra migrations/ https://academy.datastax.com/planet-cassandra/mysql-to-cassandra-migration. Accessed 10 May 2020

12. Neo4j: Import: RDBMS to Graph. https://neo4j.com/developer/relational-to-graph-import. Accessed 10 May 2020

13. Redisco 0.1.4, Python Containers and Simple Models for Redis. https://pypi.org/project/redisco/. Accessed 10 May 2020

14. Chasseur, C., Li, Y., Patel, J.M.: Enabling JSON document stores in relational systems, In: WebDB, vol. 13, pp. 14–15 (2013). http://citeseerx.ist.psu.edu/viewdoc/summary?doi=10.1.1.363.7814

15. Lee, C.-H., Zheng, Y.-L.: Automatic SQL-to-NoSQL schema transformation over the MySQL and HBase databases. In: International Conference on Consumer Electronics-Taiwan, IEEE (2015). https://doi.org/10.1109/ICCE-TW.2015.7216979

16. Hanine, M., Bendarag, A., Boutkhoum, O.: Data migration methodology from relational to NoSQL databases, world academy of science, engineering, and technology. Int. J. Comput. Elect. Autom. Control Inf. Eng. **9**, 12 (2016). https://doi.org/10.5281/zenodo.1339211

17. Oliveira, F., Oliveira, A., Alturas, B.: Migration of relational databases to NoSQL-methods of analysis, Mediterr. J. Soc. Sci. **9**(2) (2018). https://doi.org/10.2478/mjss-2018-0042

18. Sullivan, D.: NoSQL for Mere Mortals, 1st edn. Pearson Education, Boston (2015)

19. Redislabs, Redis Documentation. https://redis.io/documentation. Accessed 11 May 2020
20. Memcached, Documentation. https://memcached.org. Accessed 12 May 2020
21. Riak, Riak Products. https://riak.com/products/. Accessed 15 May 2020
22. Apache Cassandra: Apache Cassandra Documentation v4.0-alpha3. http://cassandra.apache.org/doc/. Accessed 15 Mar 2020
23. Apache HBase: Welcome to Apache HBase. https://hbase.apache.org/. Accessed 15 Mar 2020
24. Scylla: The World's Fastest NoSQL Database. https://www.scylladb.com/product/. Accessed 17 Mar 2020
25. Apache CouchDB: Apache CouchDB® 2.3.1 Documentation, http://docs.couchdb.org/en/stable/. Accessed 17 Mar 2020
26. MongoDB, Introduction to MongoDB. https://www.mongodb.com/what-is-mongodb. Accessed 17 Mar 2020
27. Amazon DynamoDB, Amazon DynamoDB Fast and flexible NoSQL database service for any scale. https://aws.amazon.com/dynamodb/. Accessed 17 Mar 2020.
28. Neo4j, The Internet-Scale Graph Platform. https://neo4j.com/product/. Accessed 17 Mar 2020
29. AllegroGraph: AllegroGraph - The Enterprise Knowledge Graph. https://allegrograph.com/. Accessed 10 Mar 2020
30. Couchbase: Data Model – Mapping from MySQL to CouchbaseServer. https://docs.couchbase.com/server/4.1/install/migrate-mysql.html. Accessed 10 Mar 2020
31. Lee, Y.S.: Near zero downtime migration from MySQL to DynamoDB, 18 April 2018. https://aws.amazon.com/blogs/big-data/near-zero-downtime-migration-from-mysql-to-dynamodb/. Accessed 8 Mar 2020
32. Riak: Moving from MySQL to Riak, 23 April 2014. https://riak.com/posts/technical/moving-from-mysql-to-riak/. Accessed 07 Mar 2020
33. MongoDB: Flexible enough to fit any industry. https://www.mongodb.com/who-uses-mongodb
34. Neo4j: The World of Graphs Powered by Neo4j. https://neo4j.com/customers/. Accessed 10 Feb 2020
35. Apache Cassandra: MySQL to Cassandra Migration. https://academy.datastax.com/planet-cassandra/mysql-to-cassandra-migration. Accessed 10 Feb 2020
36. DB engines: DB-Engine Ranking. https://db-engines.com/en/ranking. Accessed 1 Feb 2020
37. Schreiner, G.A., Duarte, D., Mello, R.d.S.: When relational-based applications go to nosql databases: a survey. Information **10**(7), 241 (2019). https://doi.org/10.3390/info10070241
38. Abadi, D., et al.: The Seattle report on database research. ACM SIGMOD Record. **48**(4), 44–53 (2020). https://doi.org/10.1145/3385658.3385668
39. Jia, T., Zhao, X., Wang, Z., Gong, D., Ding, G.: Model transformation and data migration from relational database to MongoDB. In: IEEE International Congress on Big Data (BigData Congress) 2016 June 27, pp. 60–67. IEEE (2016). https://doi.org/10.1109/BigDataCongress.2016.16
40. Karnitis, G., Arnicans, G.: Migration of relational database to document-oriented database: structure denormalization and data transformation. In: 7th International Conference on Computational Intelligence, Communication Systems and Networks 2015 June 3, pp. 113–118. IEEE (2015). https://doi.org/10.1109/CICSyN.2015.30
41. Wang, Y., Shah, R., Criswell, A., Pan, R., Dillig, I.: Data migration using Datalog program synthesis. Proc. VLDB Endow. **13**(7) (2020). https://doi.org/10.14778/3384345.3384350
42. Serrano, D., Stroulia, E.: From relations to multidimensional maps: a SQL-to-hbase transformation methodology. In: Proceedings of the 26th Annual International Conference on Computer Science and Software Engineering, pp. 156–165, October 2016. https://doi.org/10.5555/3049877.3049893

43. Unal, Y., Oguztuzun, H.: Migration of data from relational database to graph database. In: Proceedings of the 8th International Conference on Information Systems and Technologies, pp. 1–5 (2018). https://doi.org/10.1145/3200842.3200852
44. Fouad, T., Mohamed, B.: Model transformation from object-relational database to NoSQL column based database. In: Proceedings of the 3rd International Conference on Networking, Information Systems & Security, pp. 1–5, March 2020. https://doi.org/10.1145/3386723.3387881

Information Interchange of Web Data Resources

Cascaded Anomaly Detection with Coarse Sampling in Distributed Systems

Amelia Bădică[3] (ID), Costin Bădică[3] (ID), Marek Bolanowski[1(✉)] (ID), Stefka Fidanova[4] (ID),
Maria Ganzha[2] (ID), Stanislav Harizanov[4] (ID), Mirjana Ivanovic[5] (ID), Ivan Lirkov[4] (ID),
Marcin Paprzycki[2] (ID), Andrzej Paszkiewicz[1] (ID), and Kacper Tomczyk[1] (ID)

[1] Rzeszów University of Technology, Rzeszów, Poland
{marekb,andrzejp}@prz.edu.pl
[2] Polish Academy of Sciences, Warszawa, Poland
[3] University of Craiova, Craiova, Romania
[4] Institute of Information and Communication Technologies, Bulgarian Academy of Sciences,
Sofia, Bulgaria
[5] University of Novi Sad, Faculty of Sciences, Novi Sad, Serbia

Abstract. In this contribution, analysis of usefulness of selected parameters of a distributed information system, for early detection of anomalies in its operation, is considered. Use of statistical analysis, or machine learning (ML), can result in high computational complexity and requirement to transfer large amount of data from the monitored system's elements. This enforces monitoring of only major components (e.g., access link, key machine components, filtering of selected traffic parameters). To overcome this limitation, a model in which an arbitrary number of elements could be monitored, using microservices, is proposed. For this purpose, it is necessary to determine the sampling threshold value and the influence of sampling coarseness on the quality of anomaly detection. To validate the proposed approach, the ST4000DM000 (Disk failure) and CICIDS2017 (DDoS) datasets were used, to study effects of limiting the number of parameters and the sampling rate reduction on the detection performance of selected classic ML algorithms. Moreover, an example of microservice architecture for coarse network anomaly detection for a network node is presented.

Keywords: Anomaly detection · Anomaly prediction · Complex distributed system · Computer network management

1 Introduction

Despite improving quality of devices used in modern computer systems, failures continue to occur. Anomalous functioning can be caused by a hardware failure or be the result of an external factor, such as an attack. Not detecting the anomaly on time may result serious consequences, including large financial losses. Hence, failure prediction may reduce damages and costs of maintenance and repair. Besides, it may reduce system downtime and ensure service reliability for the end users. Anomaly detection is, typically, based on analysis of selected system parameters, during normal operation and

S. Sachdeva et al. (Eds.): BDA 2021, LNCS 13167, pp. 181–200, 2022.
https://doi.org/10.1007/978-3-030-96600-3_13

during failures. It is a well-known fact that in production systems, occurrence of an anomaly can be conceptualized mainly in two scenarios. (1) Anomaly detection sub-system informs about an adverse event that is "in progress", or that has ended, and the goal of administrator(s) is to limit impact of its (possible/future) effects on system's operation (e.g., DDoS detection). (2) Anomaly prediction informs of the deteriorating operational parameters, suggesting the system failure. Here, it may be possible to react in time to avoid occurrence of an actual failure (e.g., hard disk drive crash). However, in what follows, unless explicitly stated, anomaly detection and anomaly prediction are treated similarly, and terms are used interchangeably. This is because the focus of this work is on the detection itself, rather than its context.

Majority of research on anomaly detection is based on application of Machine Learning (ML, hereafter) techniques. Here, ML is applied to data collected within the observed systems. However, as it will be shown in Sect. 3, due to the amount of generated data, in most cases, only key components of large systems can be monitored. This is why, a novel method for system monitoring and anomaly detection (cascading anomaly detection with coarse sampling) has been proposed. It allows efficient monitoring of a substantially larger number of system components, facilitates distributed probing, and reduces the overall size of the analyzed dataset.

In this context, we proceed as follows. Section 2 contains summary of the state of the art. While it covers a relatively broad area of anomaly detection research, this contribution is focused on anomaly prediction in distributed computer systems. Architecture of the proposed approach is discussed in Sect. 3. Particularly important for the proposed approach are effects of limiting sampling frequency on the effectiveness of fault detection (considered in Sect. 4). This is particularly important in the case of IoT ecosystems, which are often characterized by limited bandwidth of communication channels. Section 5 summarizes the results, indicating fields of their potential implementation, and directions of future research.

2   Related Literature

Due to the scale of distributed systems, it is practically impossible to entirely rely on human-based management. Instead, automatic recognition and prediction of anomalies has to be applied [1]. For the failure prediction, algorithms based on supervised learning are mainly used. Here, in recent years, deep learning-based approaches received considerable attention [2]. Many studies have noted that certain anomalies manifest themselves as deviations of specific performance parameters. In this context, Williams et al. [3] studied which performance metrics are affected by different types of faults. These parameters included CPU usage, available memory, and network traffic, among others. On the basis of collected data, system can learn to distinguish between correct behavior and anomalies. Unfortunately, this solution does not provide information about the cause of failure and does not "confirm" failure occurrence. It only delivers "preliminary information" about possible problems. A similar approach can be found in [4], focusing on multilayer distributed systems. A solution, called PreMiSE, combines anomaly-based and signature-based techniques to identify failures. Here, it is possible to distinguish, which anomalous situations are "correct" and can lead to a failure. Based on historical data,

PreMiSE learns to associate the type of failure with specific malicious anomalies. In [5], recurrent neural networks were used to predict task failures in cloud systems, and prediction accuracy of 84% was reported. Work of Zhao et al. [6], attempted to address difficulty of describing large datasets from, for example, cloud data centers. To avoid this problem, the authors proposed a system based on k-nearest-neighbor (k-NN) clustering. Using principal component analysis [7], they extracted the most relevant features and reduced the dimension of the data matrix. Next, k-NN and the modular community detection were applied to find the largest cluster, which was assumed to represent the normal functioning, while other clusters were claimed to represent anomalies.

A large number of anomaly detection systems were applied to clouds, or to core transmission links. Here, increasing computational efficiency of detection systems allows for increasing the number of end devices and their components to be analyzed. One of the topics, emerging in recent research, is prediction of computer hardware failures [8]. Of course, due to variety of failures that may occur, ML algorithms cannot predict every failure. To use them effectively, it is necessary to select pertinent indicators that can inform about the symptoms that precede the actual failure of a selected class of components (e.g. database servers). One of the most important components, with a high probability of failure is the hard drive [9]. Exact prediction of time of failure of the selected hard drive is extremely difficult. Since, for all practical purposes, such precise prediction is not necessary, the concept of time window, indicating the approximate time of failure, was introduced. This allows early reaction, mitigating core dangers. One of the proposed approaches is to build a statistical model, consisting of several environmental and operational parameters [10]. Here, the influence of individual indicators is determined, and dependencies between them are identified. In this context, recurrent long-short term memory (LSTM) neural networks, applied for time series prediction, have been used in [11]. Here, authors managed to predict disk failures, in the next 15 days time-window, with an accuracy of 86.31%. On the other hand, system using XGBoost, LSTM and ensemble learning, achieved prediction accuracy of 78% for a 42 day window [12]. Bothe these results are reasonable for production systems.

Fault prediction materializes also in power systems, where faults lead to severe problems and power outages [13]. An approach to solving power disruption problem was proposed by Omran et al. [14]. Here, authors used various ML techniques to achieve the highest possible prediction accuracy. Their proposed system consists of data cleaning, conversion, and classification, using selected features. Applying an ant colony optimization, the most important features of the dataset are selected. Next, a number of classifiers, such as k-NN, Artificial Neural Networks, Decision Tree (DT), Logistic Regression, and Naïve Bayes (NB) were used. Performed experiments resulted in accuracy between 75% and 86%.

A different approach to the problem has been found in works dealing with power cable faults [15, 16]. As a solution, statistical analysis of data collected over a long time period, consisting of various parameters that describe network load, and environmental influence, was used.

Industrial Internet of Things (IoT) environments are another area where fault prediction is important. Due to the (very) large number of sensors used to check the status of machines and manufacturing processes, huge amount of data is collected.

Therefore, it is important to use an appropriate feature selection algorithm, to discard data that has no impact on fault prediction. Study by Kwon et al. [17], proposed a fault prediction model with an iterative feature selection. First, data is prepared by splitting, elimination, and normalization. Next, using random forest algorithm, the importance of each feature is established, based on its correlation with failure(s). Then, different sets of features are iteratively selected, and models are built on their basis, using the Support Vector Machine (SVM) classifier. Finally, model selection is performed on the basis of prediction accuracies. In their work, Fernandes et al. [18] focused on prediction of heating equipment failure. An IoT platform that provides sensor information was used as a data source. Supervised learning was applied to find classes that describe the fault condition. Feature selection, based on a random forest, was used to discard irrelevant attributes. The LSTM neural network was applied to the time series.

Another method, appearing in various publications, related to failure prediction, is the Bayesian network [19]. An example of its application is software fault prediction [20]. Besides, Bayesian networks have been used for anomaly detection, diagnosis, or time series prediction.

One more area, where anomaly detection has been applied, are computer networks [21–23]. Here, popular methods are based on mining error logs, and analysis of copies of network traffic samples. In the work of Zhong et al. [24], the investigation of data from alarm logs of a metropolitan area network, over a period of 14 months, was carried for classification of the state of the operation of the system (normal, abnormal, failure, and sudden failure). In [25], similarly as in above mentioned research, data is derived from logs. However, this time, a simulated wireless telecommunication system is considered. As a result of unstructured nature of the data, it had to be processed and labeled. A convolutional neural network model was used to make predictions about the state of the network.

3   Architecture of the Proposed System

Majority of the approaches described above are applied to monitoring only the key elements of a system. Here, a new architecture for anomaly detection and prediction in distributed systems, which can include majority of its functional elements, is proposed. To achieve this, cascade model, with coarse sampling, is to be used. A scheme of the model is shown in Fig. 1a. At each level (cascade), a set of different tools for anomaly detection is to be implemented, according to the following principle: the first cascade uses the least accurate methods with low computational complexity, capable of processing (very) large amount of data (very fast); in next cascade, level of expected accuracy of anomaly detection increases. However, here only events that have been "flagged" by the first level of the cascade are considered. Hence, the amount of data that is (has to be) taken into account is considerably smaller. The last cascade "decides" whether the received data indicates an occurrence of an actual anomaly. Overall, if an anomaly is detected as possible, at a given level, pertinent data is passed to the next cascade for further (more detailed; resource consuming) analysis. This approach allows monitoring a wider range of (non-key) elements of the system. Of course, data from critical elements (e.g., production servers) may be routed directly to the "most accurate" cascade.

Here, it is crucial to capture problems "as they materialize", without waiting for the cascading system to "capture them". In this contribution, a dual cascade system is considered. Moreover, the reported research is focused on optimizing the performance of the first cascade.

Separately, it has to be stressed that in anomaly detection, the key issue is to minimize the impact of the detection mechanisms on the functioning of individual elements of the system. To meet this requirement, additional probes may be installed. These constitute a subsystem independent from the monitored system, which can monitor, for instance, network core connections or Internet service provider access links, using specialized out-of-band intrusion detection (IDS) and intrusion prevention (IPS) subsystems. However, this approach has two major drawbacks: (1) it sends large amount of data to the centralized detection system(s), and (2) high performance of these systems needs to be guaranteed. Poor performance of IDS/IPS can introduce additional latency, which may be unacceptable. Here, work [26] proposes native mechanisms to copy traffic from selected switches and forward it for analysis outside of the production transmission path. This approach eliminates the problem of delays introduced by the anomaly detection process. It can also be successfully applied to monitor other components of a distributed system (not only network devices).

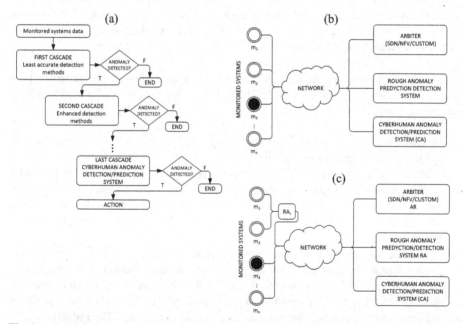

Fig. 1. a. Model of a cascaded anomaly detection system; **b.** Centralized coarse anomaly detection system; **c.** Distributed anomaly detection system.

In this context, the scheme of the centralized system with coarse anomaly detection is presented in Fig. 1b. The monitored systems m_1, m_2 ... m_n (network devices, computer systems, printers, IoT components, etc.) send diagnostic information via a (possibly separate) network to the supervisory systems, i.e.:

- **Rough anomaly prediction/detection (RA) system** – which is responsible for analysis of large sets of diagnostic data, from all supervised devices. RA systems can be implemented in the form of hardware-software probes, which are inserted in selected places within the monitored system. The detection algorithms used in RAs should be characterized by lowest possible complexity, and a minimum acceptable detection threshold (configurable by the administrator). When a potential anomaly is detected, pertinent data is forwarded to the CA system.
- **Cyber-Human anomaly prediction/detection system (CA)** – is a collection of IDS and IPS systems, which uses highly accurate (though computationally complex) detection systems. Its results are to be verified by the engineering team. Note that in critical systems, such as production servers, it may be necessary to send full diagnostic data directly to the CA, bypassing the RA. Observe also, that in multi-step cascade systems, it is possible to instantiate (in selected locations) additional RA layers, with "intermediate level of sensitivity" of anomaly detection.
- **Arbiter (AR)** – system for control and management of monitored devices, which, based on data from the anomaly detection systems, determines parameters of the monitored system and an identified event. On this basis, it may undertake reconfiguration actions, e.g.: disabling a device, blocking it using access control lists, reconfiguring routing, sending a service request, performing a backup, etc. The AR architecture can be built based on the SDN (software-defined networking) model, NFV (network functions virtualization), Ansible, or by using custom solutions written in Python, with the Netmiko and Paramiko libraries [27–29].

As noted, in case of large ecosystems, collecting data from all supervised devices puts a significant load on the network. The following approaches can be used to reduce the network load:

- Monitoring only key system elements (e.g. m_3, marked in black), is a common approach, which obscures the complete picture of the state of the system. As a side note observe that deficit of ICT engineers makes it increasingly difficult to ensure regular review of non-key IT systems (e.g. disks in office PCs). Hence, only by introducing automatic supervision of all (as many as possible) systems, it is possible to allow prediction of failures on non-core components and undertaking timely remedial actions.
- Relocating very coarse detection systems, e.g. RA_L (see, Fig. 1c), where the data is generated (e.g. in the three cascade approach). This reduces transfer of diagnostic data across the network. Data is sent to RA and AR only when potential anomalies are detected. The RA_Ls, operating within a single system, may differ in applied classifiers and their settings, to match specific classes of monitored devices. Their relatively simple structure allows them to be implemented as microservices, and deployed anywhere in the distributed system, e.g. on servers, IoT devices, gateways, PCs, etc. Moreover, RA_Ls, as the first cascade step, can be dedicated to non-key devices, where high accuracy of threat detection may not be required. This also reduces the complexity of the detection process (for such systems) and reduces the total number of RA_L probes in the system.

- Using an appropriate (reduced) sampling rate for diagnostic data sent to RA and/or RA_Ls, to reduce the data transfer. Moreover, using only "key parameters" to train the ML model (while omitting the less "important ones") can also lead to faster model training and application. Here, in both cases, it is necessary to determine what effect removal of some available data has on the accuracy of anomaly detection. This aspects of the proposed approach is considered in detail in Sect. 4.

4   Effect of Number of Parameters and Sampling Frequency on Model Accuracy

To establish relation between the number of parameters and the number of samples and the quality of anomaly detection, data from two, publicly available, datasets was used:

- ST4000DM000 dataset (from 2017) was selected from the Blackblaze website [31]. It contains statistics of hard drives from data centers. Each row of data refers to a single day, and contains, among others, the following attributes: time, disk serial number, model, capacity, and statistics derived from the S.M.A.R.T. protocol. Additional column, named "Failure" having value "0" if the disk is operational, or "1" if it is the last day before failure, is also present. Table 1 shows the number of records assigned to each class describing disk performance.
- CICIDS2017 [32] dataset contains 72 attributes of network traffic, during normal activities and during various types of attacks. Here, we have taken into consideration data related to a DDoS attack. Data was collected during five days, in the form of pcap's and event log files. Additional column, named "Label", having a value of "0" if there is no DDoS attack and "1" when packets are part of the DDoS attack, was also included (Table 2).

Table 1. Class types and record counts in dataset ST4000DM000

Normal ('0')	Failure ('1')	Total number of records
167823 (94,1%)	10477 (5,9%)	178300

Table 2. Class types and record counts in dataset CICIDS2017

Normal ('0')	Failure ('1')	Total number of records
77585 (26%)	221485 (74%)	299070

In this work, which should be treated as a preliminary step of a long-term project, four classic ML algorithms were experimented with: SVM, DT, k-NN and NB. In the future, we plan to look into pertinent modern approaches, e.g. encoders, LSTM and CNN neural networks, etc. Models were implemented in MATLAB, with the "Statistics and Machine

Learning Toolbox" add-on [30]. Standard versions of the following measures were used to compare model performance: accuracy, precision, recall, specificity, and F-measure. Before proceeding, let us recall (and stress) that at this junction we are interested in methods that are to be applied in the first cascade. Hence, they should be "relatively accurate", while consuming minimal amount of resources, and being fast in training and application.

4.1 Reducing Number of Attributes

In the first series of experiments, effects of reducing the number of attributes were studied. To determine the order of adding attributes, the MRMR (minimum redundancy maximum relevance) algorithm was used, which returns attributes in order of importance to the classification. Accordingly, attributes were added starting from these that were established to be the most relevant. Figure 2, Fig. 3, Fig. 4 and Fig. 5 show results of the DT, SVM, k-NN and NB, for the dataset ST4000DM000, for different number of attributes, added in the following order (we list the first 10): Current Pending Sector Count; Reported Uncorrectable Errors; Power-off Retract Count; Reallocated Sectors Count; Uncorrectable Sector Count; End-to-End error; UltraDMA CRC Error Count; Temperature; Total LBAs Read; Start/Stop Count.

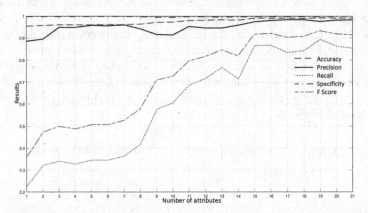

Fig. 2. Effect of varying the number of attributes for the DT model; ST4000DM000 dataset.

In case of DT (Fig. 2) particularly noticeable is the change in the recall measure responsible for the model's ability to detect abnormal disk behavior. It can be concluded that the DT model needs most of available attributes to get a satisfactory result. For the SVM algorithm, a similar relationship was observed (see Fig. 3). However, for this algorithm, addition of the last two attributes resulted in a significant decrease in both F-Score and Recall.

With the k-NN algorithm (Fig. 4), high sensitivity of performance, to successively added initial (most important, according to the MRMR) attributes was observed. As the number of attributes increased, the performance measures oscillated, until the number of attributes reached 15, where they stabilized at very high values. Interestingly, again, adding the last attribute had a negative impact on the results.

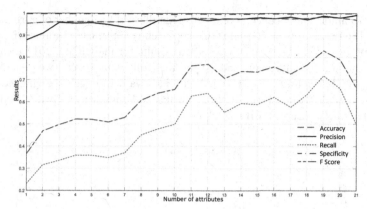

Fig. 3. Effect of varying the number of attributes on SVM model; ST4000DM000 dataset.

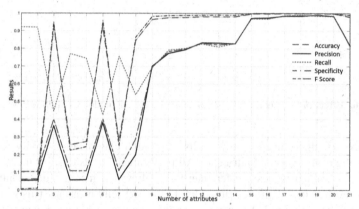

Fig. 4. Effect of varying the number of attributes on k-NN model; ST4000DM000 dataset.

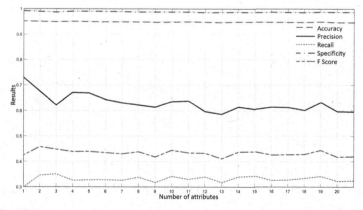

Fig. 5. Effect of varying the number of attributes on NB model; ST4000DM000 dataset.

The NB algorithm obtained poor results regardless of the number of attributes used (Fig. 5). The results show clearly that this approach is not suitable for this type of data.

Figure 6, Fig. 7, Fig. 8 and Fig. 9 show results for the CICIDS2017 dataset, for different number of attributes, added in the following order (we list the first 10): Init_Win_bytes_forward, Init_Win_bytes_backward, MinPacketLenght, ActiveStd, BwdPacketLenghtMean, FwdIATMin, FwdPacketLenghtMax, BwdPacketLenght-Mean, IdleStd, TotalLenghtOfBwdPackets.

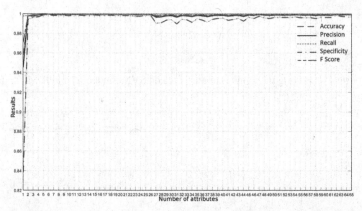

Fig. 6. Effect of varying the number of attributes on DT model; CICIDS2017 dataset.

For the DT algorithm (Fig. 6), results show high classification ability for as few as two of the most important attributes. All performance measures, for more than two parameters, obtained values greater than 98%. However, increase of the number of attributes resulted in a slight deterioration of the specificity measure. This would seem to suggest that DT, with 2–4 attributes should is a reasonable candidate for the first cascade step in the proposed approach to anomaly detection.

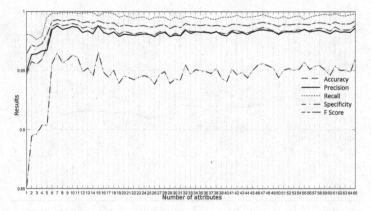

Fig. 7. Effect of varying the number of attributes on SVM model; CICIDS2017 dataset.

Figure 7 shows the result for the SVM algorithm. Interestingly, the best performance was obtained for 6–15 attributes. There exists some level of similarity to the results, obtained by the SVM, for the ST4000DM000 dataset. The results obtained for the Specificity measure remain very low regardless of the number of included parameters.

Figure 8 shows results of the model dependence on the number of attributes for k-NN algorithm and the CICIDS2017 dataset. Here, the results are rather peculiar as sudden performance jump can be observed after 35th parameter is added. From there on, very high performance of all measures is observed. It is unclear why this specific parameter brought this change. However, taking into account that the DT approach reaches high level of accuracy (for all measures) already for very few parameters, we have decided to not to investigate this peculiar effect further.

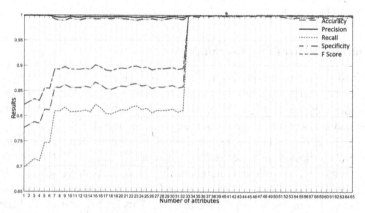

Fig. 8. Effect of varying the number of attributes on k-NN model; CICIDS2017 dataset.

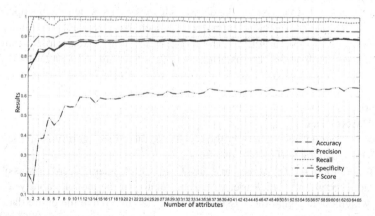

Fig. 9. Effect of varying the number of attributes on NB model; CICIDS2017 dataset.

Figure 9 shows the results for the NB algorithm and the CICIDS2017 dataset. The figure shows a significant increase in measure values at the initial stage of attribute addition. Already at about half of the added attributes, the models gains its maximum performance, which remains stable. However, again, also for this dataset, the performance of NB is substantially worse than that of the remaining three classic algorithms.

Overall, taking into account the results obtained when investigating effects of reducing the number of analyzed attributes; for the prediction of disk damage the best results were obtained after application of the k-NN algorithm, for 16–20 attributes. In the case of problems related to the detection of anomalies for network traffic, the best turned out to be the DT algorithm, which obtained very good results already for two attributes. These results indicate clearly, that anomaly detection, performed by the first step of the cascade may really need to use different ML approaches, when anomalies in different contexts are to be detected.

4.2 Reducing the Size of the Learning Set

In the second series of experiments, the effect of granularity of data sampling, on the performance of predictive models was explored. In these experiments, the same performance metrics were applied. This allowed us to observe how much the ML process can be accelerated at the cost of a small decrease in model accuracy.

Initially, it was decided to use a systematic sampling approach. In this approach, we select every $2n$th element in the entire set, where $n \in \{1, 2, \ldots, 13\}$. For example, for $n = 1$ elements with the following numbers: $1, 2, 4, 6, 8, \ldots$ were selected. However, for the ST4000DM000 dataset, a large class imbalance was observed, with the number records in class 1 being only 5.9% of the total set. This led to a situation where smaller data samples contained records belonging only to class 0, which made it impossible to achieve valid results. Therefore, simple random sampling of the dataset was used (to assure existence of both classes in the dataset. However, this issue remains open for the case of practical realization of the proposed approach. Not being able to collect enough data samples of Class 1 to initialize the model needs to be taken into account. Here, the potential of application of transfer learning (e.g. using the ST4000DM000 dataset) may need to be considered. In addition, it was also decided to use 10 times repetition of the test, to calculate the average value for the obtained measures. Figure 10 shows a comparison of results for systematic and random sampling for the DT algorithm on ST4000DM000 data.

Figure 11 through Fig. 16 show the results for the other algorithms applied on the ST4000DM000 and CICIDS2017 datasets.

For the ST4000DM000 dataset, and the DT algorithm, a large decrease was observed in precision, recall and F-measure, which are related to the model's ability to predict class 1, which could signal the possibility of disk failure. When the entire dataset was analyzed before sampling, the values of these parameters were close to 100%, while for 10% of the input data, the values dropped to about 78%.

For the CICIDS2017 dataset, the results for the DT algorithm are shown in Fig. 12. Here, a fairly slow decrease in the performance measures can be observed. The largest drop was observed for the specificity measure.

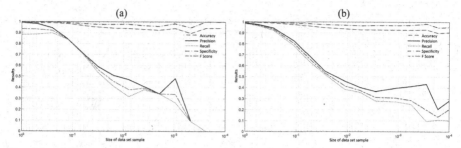

Fig. 10. Comparison of systematic (a) and random (b) sampling for the ST4000DM000 dataset for the DT algorithm.

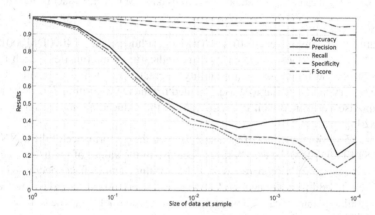

Fig. 11. Effect of changing the number of records on DT model; ST4000DM000 dataset.

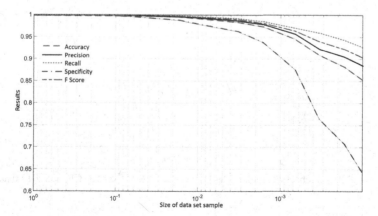

Fig. 12. Effect of changing the number of records on the DT model; CICIDS2017 dataset.

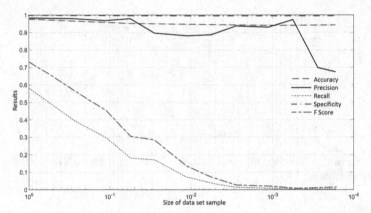

Fig. 13. Effect of changing the number of records on SVM model; ST4000DM000 dataset.

Figure 13 contains the results for the SVM algorithm and the ST4000DM000 dataset. Due to the small number of class 1 records (indicating a malfunction), each time the set is reduced, the failure prediction ability is greatly affected. The results shown in Fig. 14 are very much similar to those obtained for the DT algorithm. Overall, the SVM algorithm also performed well in prediction for the reduced number of records of the CICIDS2017 set.

Figure 15 shows the effect of data sampling on the performance of the k-NN model and the ST4000DM000 dataset. A rapid decrease in the values of precision, recall and F-Score were observed. Compared to the DT algorithm, a much faster decrease in values can be seen. In addition, the accuracy and specificity measures record a slow decrease towards results that are meaningless (form the perspective of these measures).

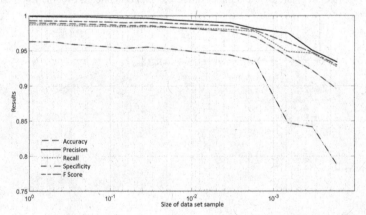

Fig. 14. Effect of changing the number of records on SVM model; CICIDS2017 dataset.

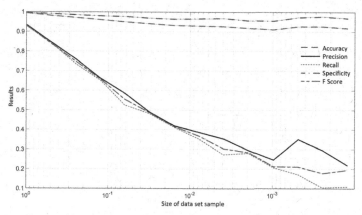

Fig. 15. Effect of changing the number of records on k-NN model; ST4000DM000 dataset.

For the k-NN learning model and the CICIDS2017 dataset (Fig. 16), similar results as for the DT algorithm are observed. Only when the learning set is significantly reduced, there is a greater decrease in the performance measures. The largest performance drop is observed for the specificity measure.

For the NB algorithm, the use of data sampling proved to be problematic due to strong assumptions about the learning datasets, leading to problems with missing variance, for some of the attributes. Due to this error, data sampling was performed to the maximum possible value. Very unsatisfactory results were obtained for the ST4000DM000 set (Fig. 17). The accuracy and specificity remain fairly constant while the model is unable to correctly classify of the possible failures due to the value of the sensitivity measure.

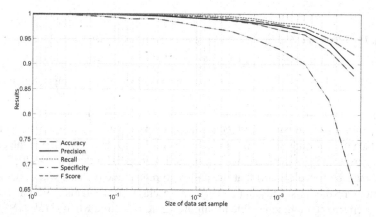

Fig. 16. Effect of changing the number of records on k-NN model; CICIDS2017 dataset

For the CICIDS2017 data, the NB algorithm keeps the performance measure values constant, for the reduced amount of learning data (Fig. 18). The largest increase was recorded for the specificity measure for a learning set size below 10% of the original size when at the same time the other measures recorded a small increase.

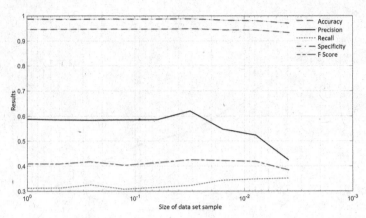

Fig. 17. Effect of changing the number of records on NB model; ST4000DM000 dataset.

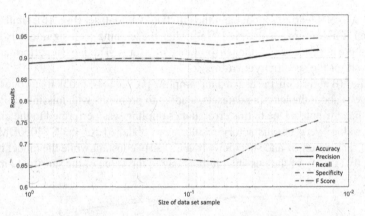

Fig. 18. Effect of changing the number of records on NB model; CICIDS2017 dataset.

The obtained results show that it is possible to reduce the sampling frequency of the learning set data, while maintaining threshold that would be satisfactory for detection and prediction of anomalies, when applied within the first step of the cascade. By using this approach, it is possible not only to reduce the amount of data sent over the networks, but also reduce the learning time. We illustrate this in Fig. 19. Here, we depict the effect of sampling frequency on the learning time of the DT model, for the CICIDS2017 dataset.

It should be noted (again) that the ability to reduce the dataset, while maintaining satisfactory results is strictly dependent on the class of problems. The CICIDS2017 dataset seems to be more amenable to sampling rate reduction when DT, k-NN or SVM algorithms are used. For the ST4000DM00 dataset, the best results are obtained for the k-NN algorithm, but the susceptibility of this dataset to reduce the learning set, while maintaining acceptable failure detection results is low compared to the results obtained for the CICIDS2017 dataset.

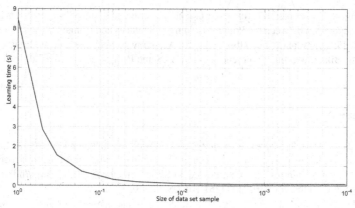

Fig. 19. Effect of changing the number of records on the learning time of the DT model; CICIDS2017 dataset.

5   Concluding Remarks

The results presented in this work represent a contribution to the development of a novel architecture for distributed anomaly detection and prediction in a distributed system. The proposed solution is based on a cascade model, in which the first detection step can be applied, for instance, to non-key systems, which often are not monitored. Classic ML algorithms were applied for anomaly detection. Experimental work was focused on studying effects of reducing (1) number of measured attributes, and (2) data sampling rate, on the efficiency of threat detection by the first level of detectors (RA systems). Obtained results show that it is possible to use limited datasets without significant degradation of anomaly detection performance. However, a significant influence on the results has matching of applied algorithms with the monitored element. Moreover, the specifics of the monitored system are going to determine the minimum acceptable threshold for anomaly detection. Achieved results are rather promising, however, it is necessary to further continue research in this area, using more datasets and more recently proposed ML approaches in order to further experimentally ground the proposed approach and to provide guidelines for its applicability.

In this context, the proposed solution will be implemented and tried in a production system, with large number of disks and auxiliary systems. The work will also focus on developing effective mechanisms for automatic management and reconfiguration of distributed system elements in response to anomalies detected by the proposed approach. Here, preventive measures applicable at each step of the cascade, for different contexts will be formulated.

Acknowledgement. This project is financed by the Minister of Education and Science of the Republic of Poland within the "Regional Initiative of Excellence" program for years 2019–2022. Project number 027/RID/2018/19, amount granted 11 999 900 PLN.

This work has been supported by the joint research project "Agent Technologies in Dynamics Environments" under the agreement on scientific cooperation between University of Novi Sad, University of Craiova, SRI PAS and Warsaw University of Technology, as well as by the joint research project "Novel methods for development of distributed systems" under the agreement on scientific cooperation between the Polish Academy of Sciences and Romanian Academy for years 2019–2021. Finally, support from the Bulgarian Academy of Sciences and the Polish Academy of Sciences, (Bilateral grant agreement between BAS and PAS) is acknowledged.

References

1. Janus, P., Ganzha, M., Bicki, A., Paprzycki, M.: Applying machine learning to study infrastructure anomalies in a mid-size data center - preliminary considerations. In: Proceedings of the 54th Hawaii International Conference on System Sciences (2021). http://hdl.handle.net/10125/70636. https://doi.org/10.24251/HICSS.2021.025

2. Neu, D.A., Lahann, J., Fettke, P.: A systematic literature review on state-of-the-art deep learning methods for process prediction. Artif. Intell. Rev. (2021). https://doi.org/10.1007/s10462-021-09960-8

3. Williams, A.W., Pertet, S.M., Narasimhan, P.: Tiresias: black-box failure prediction in distributed systems. In: Proceedings of the 2007 IEEE International Parallel and Distributed Processing Symposium, Long Beach, CA, USA, pp. 1–8. IEEE (2007). https://doi.org/10.1109/IPDPS.2007.370345

4. Mariani, L., Pezzè, M., Riganelli, O., Xin, R.: Predicting failures in multi-tier distributed systems. J. Syst. Softw. **161**, 110464 (2020). https://doi.org/10.1016/j.jss.2019.110464

5. Chen, X., Lu, C., Pattabiraman, K.: Failure prediction of jobs in compute clouds: a google cluster case study. In: Proceedings of the 2014 IEEE International Symposium on Software Reliability Engineering Workshops, Naples, Italy, pp. 341–346. IEEE (2014). https://doi.org/10.1109/ISSREW.2014.105

6. Zhao, J., Ding, Y., Zhai, Y., Jiang, Y., Zhai, Y., Hu, M.: Explore unlabeled big data learning to online failure prediction in safety-aware cloud environment. J. Parallel Distrib. Comput. **153**, 53–63 (2021). https://doi.org/10.1016/j.jpdc.2021.02.025

7. https://medium.com/apprentice-journal/pca-application-in-machine-learning-4827c07a61db. Accessed 18 Nov 2021

8. Chigurupati, A., Thibaux, R., Lassar, N.: Predicting hardware failure using machine learning. In: Proceedings of the 2016 Annual Reliability and Maintainability Symposium (RAMS), Tucson, AZ, USA, pp. 1–6 (2016). https://doi.org/10.1109/RAMS.2016.7448033

9. Suchatpong, T., Bhumkittipich, K.: Hard Disk Drive failure mode prediction based on industrial standard using decision tree learning. In: Proceedings of the 2014 11th International Conference on Electrical Engineering/Electronics, Computer, Telecommunications and Information Technology (ECTI-CON), Nakhon Ratchasima, Thailand, pp. 1–4. IEEE (2014). https://doi.org/10.1109/ECTICon.2014.6839839

10. Strom, B.D., Lee, S.C., Tyndall, G.W., Khurshudov, A.: Hard disk drive reliability modeling and failure prediction. In: Proceedings of the Asia-Pacific Magnetic Recording Conference 2006, Singapore, pp. 1–2. IEEE (2006). https://doi.org/10.1109/APMRC.2006.365900

11. Hu, L., Han, L., Xu, Z., Jiang, T., Qi, H.: A disk failure prediction method based on LSTM network due to its individual specificity. Proc. Comput. Sci. **176**, 791–799 (2020). https://doi.org/10.1016/j.procs.2020.09.074

12. Li, Q., Li, H., Zhang, K.: Prediction of HDD failures by ensemble learning. In: Proceedings of the 2019 IEEE 10th International Conference on Software Engineering and Service Science (ICSESS), Beijing, China, pp. 237–240. IEEE (2019). https://doi.org/10.1109/ICSESS47205.2019.9040739

13. Zhang, S., Wang, Y., Liu, M., Bao, Z.: Data-based line trip fault prediction in power systems using LSTM networks and SVM. IEEE Access **6**, 7675–7686 (2018). https://doi.org/10.1109/ACCESS.2017.2785763

14. Omran, S., El Houby, E.M.F.: Prediction of electrical power disturbances using machine learning techniques. J. Amb. Intel. Hum. Comput. **11**(7), 2987–3003 (2019). https://doi.org/10.1007/s12652-019-01440-w

15. Mehlo, N.A., Pretorius, J.H.C., Rhyn, P.V.: Reliability assessment of medium voltage underground cable network using a failure prediction method. In: Proceedings of the 2019 IEEE PES Asia-Pacific Power and Energy Engineering Conference (APPEEC), Macao, China, pp. 1–5. IEEE (2019). https://doi.org/10.1109/APPEEC45492.2019.8994720

16. Sachan, S., Zhou, C., Bevan G., Alkali, B.: Failure prediction of power cables using failure history and operational condition. In: Procedings of the 2015 IEEE 11th International Conference on the Properties and Applications of Dielectric Materials (ICPADM), Sydney, NSW, Australia, pp. 380–383. IEEE (2015). https://doi.org/10.1109/ICPADM.2015.7295288

17. Kwon, J.-H., Kim, E.-J.: Failure prediction model using iterative feature selection for industrial internet of things. Symmetry **12**, 454 (2020). https://doi.org/10.3390/sym12030454

18. Fernandes, S., Antunes, M., Santiago, A.R., Barraca, J.P., Gomes, D., Aguiar, R.L.: Forecasting appliances failures: a machine-learning approach to predictive maintenance. Information **11**, 208 (2020). https://doi.org/10.3390/info11040208

19. Cai, Z., Sun, S., Si, S., Wang, N.: Research of failure prediction Bayesian network model. In: Proceedings of the 2009 16th International Conference on Industrial Engineering and Engineering Management, Beijing, China, pp. 2021–2025. IEEE (2009). https://doi.org/10.1016/10.1109/ICIEEM.2009.5344265

20. Bai, C.G., Hu, Q.P., Xie, M., Ng, S.H.: Software failure prediction based on a Markov Bayesian network model. J. Syst. Softw. **74**(3), 275–282 (2005). https://doi.org/10.1016/j.jss.2004.02.028

21. Bhuyan, M.H., Bhattacharyya, D.K., Kalita, J.K.: Network Traffic Anomaly Detection and Prevention: Concepts, Techniques, and Tools. Springer, Cham (2018). https://doi.org/10.1007/978-3-319-65188-0. ISBN 978-3-319-87968-0

22. Bolanowski, M., Twaróg, B., Mlicki, R.: Anomalies detection in computer networks with the use of SDN. Meas. Autom. Monit. **9**(61), 443–445 (2015)

23. Bolanowski, M., Paszkiewicz, A.: The use of statistical signatures to detect anomalies in computer network. In: Gołębiowski, L., Mazur, D. (eds.) Analysis and Simulation of Electrical and Computer Systems. LNEE, vol. 324, pp. 251–260. Springer, Cham (2015). https://doi.org/10.1007/978-3-319-11248-0_19

24. Zhong, J., Guo, W., Wang, Z.: Study on network failure prediction based on alarm logs. In: Proceedings of the 2016 3rd MEC International Conference on Big Data and Smart City (ICBDSC), Muscat, Oman, pp. 1–7. IEEE (2016). https://doi.org/10.1109/ICBDSC.2016.7460337

25. Ji, W., Duan, S., Chen, R., Wang, S., Ling, Q.: A CNN-based network failure prediction method with logs. In: Proceedings of the 2018 Chinese Control and Decision Conference (CCDC), Shenyang, China, pp. 4087–4090. IEEE (2018). https://doi.org/10.1109/CCDC.2018.8407833

26. Bolanowski, M., Paszkiewicz, A., Rumak, B.: Coarse traffic classification for high-bandwidth connections in a computer network using deep learning techniques. In: Barolli, L., Yim, K., Enokido, T. (eds.) CISIS 2021. LNNS, vol. 278, pp. 131–141. Springer, Cham (2021). https://doi.org/10.1007/978-3-030-79725-6_13

27. Bolanowski, M., Paszkiewicz, A., Kwater, T., Kwiatkowski, B.: The multilayer complex network design with use of the arbiter. Monographs in Applied Informatics, Computing in Science and Technology, Wydawnictwo Uniwersytetu Rzeszowskiego, Rzeszów, pp. 116–127 (2015). ISBN 978-83-7996-140-5

28. https://pypi.org/project/netmiko. Accessed 8 Nov 2021
29. https://www.paramiko.org. Accessed 8 Nov 2021
30. https://www.mathworks.com/help/stats/getting-started-12.html. Accessed 8 Nov 2021
31. https://www.backblaze.com/b2/hard-drive-test-data.html. Accessed 8 Nov 2021
32. Sharafaldin, I., Habibi Lashkari, A., Ghorbani, A.: Toward generating a new intrusion detection dataset and intrusion traffic characterization. In: Proceedings of the 4th International Conference on Information Systems Security and Privacy - ICISSP, Funchal, Madeira, Portugal, pp. 108–116. SciTePress (2018). https://doi.org/10.5220/0006639801080116

Video Indexing System Based on Multimodal Information Extraction Using Combination of ASR and OCR

Sandeep Varma[1(✉)], Arunanshu Pandey[1(✉)], Shivam[1(✉)], Soham Das[2(✉)], and Soumya Deep Roy[2(✉)]

[1] ZS Associates, Pune, India
{sandeep.varma,arunanshu.pandey,shivam.shivam}@zs.com
[2] Department of Metallurgical and Material Engineering, Jadavpur University, Kolkata, India
sohamju22@gmail.com, sdrjumme@gmail.com

Abstract. With the ever-increasing internet penetration across the world, there has been a huge surge in the content on the worldwide web. Video has proven to be one of the most popular media. The COVID-19 pandemic has further pushed the envelope, forcing learners to turn to E-Learning platforms. In the absence of relevant descriptions of these videos, it becomes imperative to generate metadata based on the content of the video. In the current paper, an attempt has been made to index videos based on the visual and audio content of the video. The visual content is extracted using an Optical Character Recognition (OCR) on the stack of frames obtained from a video while the audio content is generated using an Automatic Speech Recognition (ASR). The OCR and ASR generated texts are combined to obtain the final description of the respective video. The dataset contains 400 videos spread across 4 genres. To quantify the accuracy of our descriptions, clustering is performed using the video description to discern between the genres of video.

Keywords: Optical Character Recognition · Automatic Speech Recognition · Video analytics · Natural language processing · K-means clustering

1 Introduction

With a large scale penetration of internet access throughout the world, there has been a tectonic shift in the way we create and consume media. In a way, the percolation of internet has facilitated the democratization of media allowing anybody with an internet connection to broadcast their voice and self-publish. This has resulted in a concomitant rise in content on the internet. Video is one of the most powerful forms of content. This can be validated by an App Annie report according to which the total time spent across video streaming platforms hit

1 trillion hours in 2020. YouTube[1] clearly dominates as the platform of choice for most consumers. According to a Statista report , around 500 h of video gets uploaded to YouTube per minute.

With COVID-19, there has been a surge in online learning in the form of tutorials and lectures. For example, many pharmaceutical companies may have hours of video content on details of drugs and vaccines, webinars of doctors etc. which they might want to integrate into a private video search. Often such companies refrain from putting out the videos on a public video streaming platform like YouTube primarily due to the sensitive and private nature of the videos. Unlike most videos on public streaming platforms which have a description with keywords, these videos seldom contain such metadata. Hence, it is imperative to generate descriptions of these videos.

A video is in essence a combination of visuals or frames and audio. Hence, information is embedded in a video in two formats. To extract content from the video, we therefore need to create separate embeddings for both visuals/frames and audio. We use Optical Character Recognition (OCR) on the frames of the video and Automatic Speech Recognition on the audio to create visual and audio embeddings respectively. These embeddings are then fused together to create a description of the video in question. A video retrieval system is then developed using the description of the videos. In order to quantify how well the fusion of visual and audio embeddings works, we use 4 genres of videos and cluster them using K-Means Clustering on the description of the video. Unlike other works enumerated in literature where a specific type of video, mostly educational lectures, is used, the present work uses a wide variety of video genres. This ensures universality and renders robustness to our model.

In a nutshell, the contributions of our proposed work are as follows:

1. Fusing the textual information obtained from OCR-generated text and audio content from ASR-generated text to create description of the video.
2. To test the accuracy of the descriptions created, K-Means clustering is done on the video descriptions to discern between the genres of video.
3. A video retrieval system using the video descriptions.
4. Using a video dataset of multiple genres and types of videos.

The remaining paper is arranged as follows: Sect. 2 details the research endeavors pertaining to Multimodal Information Extraction from videos detailed in literature. Section 3 details our proposed model along with the relevant dataset description while the experimental results are enumerated and discussed in Sect. 4. Finally, the paper concludes in Sect. 5 with the future research direction.

2   Literature Survey

Retrieval of imperative information from videos has become a necessity nowadays. Over the last few years, many research works have been carried out in

---

[1] https://www.youtube.com/.

this field. In this section, we have discussed some of these research endeavours where researchers have propounded methodologies to extract vital information from videos. Adcock et al. [2] developed and presented TalkMiner, an interface wherein one could browse through videos and search for particular keywords. OCR was employed to extract keywords from videos. Epshtein el al. [6] presented an image operator that measures the value of stroke width (SW) for an image pixel. The operator presented was local and data-driven, making it quick and robust enough to avoid the requirement for multi-scale calculation. In [10], Yang et al. proposed a fast localization-verification pipeline for video text spotting and identification. Probable text candidates with high recall rate were recognized using an edge based multiscale text detector. Refining was carried out on the detected texts utilizing an image entropy-based filter. False alarms were eliminated using SW transform and Support Vector Machine (SVM)-based verification mechanisms. In 2012, Balagopalan et al. [3] suggested a statistical methodology to automate keyphrase elimination from video lecture audio transcripts. A Naive Bayes (NB) classifier based supervised ML was employed to extract pertinent keyphrases. The extracted keyphares were used as features for topic based segmentation of the videos. Balagopalan et al. [4] extended their previous work [3] by proposing a multimodal metadata extraction that made use of NB classifier and a rule-based refiner for effectual retrieval of the metadata in a video. Features from both slide content and audio trans Jeong et al. [7] employed Scale-invariant feature transform (SIFT) and adaptive threshold in their lecture video segmentation method. It offered proper segmentation of the input video stream to fetch vital details. In 2014, Yang and Meinel [9] presented an approach for mechanized automatic indexing and video retrieval. They used automated video segmentation and keyframe identification to provide a visual guidance for video content navigation. This was followed by textual metadata extraction by employing OCR on keyframes and ASR on audio file. Keywords were obtained from OCR detected text and ASR transcript. Keywords were further used for content based video searching. The drawback of this method is that it is not compatible for open data resources. Pranali et al. [8] gave preference to in-content of a video over its title and metadata explanation. Medida and Ramani [1] Recently, Chand et al. [5] assessed the existing research works pertaining to the video content-based search system. Their survey work offered a thorough overview enumerating the obstacles, challenges and, recent upgrades in this domain.

3   Proposed Methodology

The process of information extraction employed in the current paper involves extraction of textual and audio information using OCR and ASR followed by fusion of the two, to create a description of the video. These descriptions are then used for video retrieval. The details of the pipeline are shown in the Fig. 1.

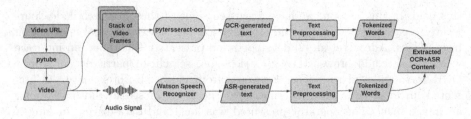

Fig. 1. Proposed pipeline

3.1   Dataset Description

The dataset has been prepared by curating videos of various duration. The in-house dataset comprises of YouTube videos chosen across 4 genres. These genres include Science, Finance and Business, Literature and Sports. From each genre, we have chosen 250 video each. Thus, in all, we have 1000 videos spread across the aforementioned genres. All the videos are downloaded from YouTube utilizing the pytube[2] library. It is noteworthy that unlike other research works, we have covered videos of various genres in order to assess the robustness of our proposed model.

3.2   Content Extraction from Videos

As described already, content extraction from videos is done by first segmenting the video into frames and audio. Textual content is then extracted using OCR on the frames. Similarly, audio content is extracted using ASR on the audio. The textual and audio data is then fused together to create the final description of the video. These parts are described in detail in the succeeding subsections.

Textual Content Extraction Using OCR. In this part, we first capture frames from the video at a fixed frame rate. This generates a stack of frames for each video. The stack of frames might contain multiple redundant and repetitive frames. Advanced image processing techniques are adopted before ingesting the images into the OCR for text extraction. Images go through a series of pre-processing methods like re-scaling, binarization, noise removal, de-skewing, blurring and, thresholding. Images are competently pre-processed to ameliorate the OCR performance. A detailed analysis is conducted to observe the accuracy of OCR as shown in Fig. 2. Hence, the text extracted using Python-tesseract (pytesseract) OCR[3] on a particular pre-processed frame is compared with the text extracted from the previous frame using cosine similarity. If the cosine similarity is greater than a certain threshold, the frame is considered redundant and the OCR generated text from that frame is rejected. The raw text generated

---

[2] https://pypi.org/project/pytube/.
[3] https://pypi.org/project/pytesseract/.

from the OCR needs to be refined and processed before it can be used as a part of description. Hence, the text is pre-processed to remove punctuation and stop words. We also perform lemmatization on the text to reduce each word to its root form. The NLP operations are executed using the NLTK library[4].

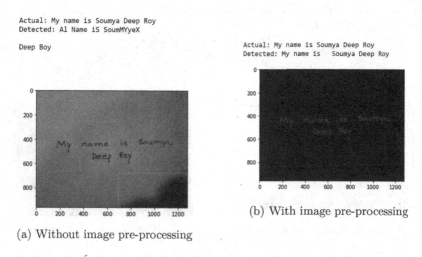

(a) Without image pre-processing

(b) With image pre-processing

Fig. 2. OCR detected texts

Audio Content Extraction Using ASR. In this part, the video is first converted to audio. The audio is then transcribed into text using ASR. ASR was implemented using IBM Watson Speech to Text API[5]. The raw audio thus obtained needs to be pre-processed before audio content extraction. Some of the pre-processing steps employed include noise removal and pre-emphasis. Ambient noise is any unwanted signal which interferes with the signal being monitored. These include sounds of electric instruments, traffic noise among others. In pre-emphasis, an audio may have frequency components which taper off at higher frequencies. Both ambient noise and pre-emphasis are rectified using relevant filters.

The processed audio thus obtained is passed through the ASR. The ASR generated raw text is processed to remove punctuations and stop words and lemmatized (Fig. 2).

3.3   Video Indexing Using OCR and ASR Generated Description

Once the description is generated by combining the OCR and ASR generated texts, we would like to search for keywords from the videos. Since the information in the video has been encoded in the form of text, the problem boils down to

---

[4] https://www.nltk.org/.
[5] https://cloud.ibm.com/apidocs/speech-to-text.

searching for keywords from a corpus of text. This task of indexing can be easily accomplished by using Whoosh[6]. Various scoring mechanisms can be used to rank the relevance of the videos. Some of the scoring methods used popularly are Frequency, TF-IDF (Term Frequency-Inverse Document Frequency), BM25F and Cosine Similarity. In the present paper, we use cosine similarity which can be enumerated by the following formula:

$$\cos(\mathbf{t}, \mathbf{e}) = \frac{\mathbf{te}}{\|\mathbf{t}\|\|\mathbf{e}\|} = \frac{\sum_{i=1}^{n} \mathbf{t}_i \mathbf{e}_i}{\sqrt{\sum_{i=1}^{n} (\mathbf{t}_i)^2}\sqrt{\sum_{i=1}^{n} (\mathbf{e}_i)^2}} \tag{1}$$

(a) Without image pre-processing   (b) With image pre-processing

Fig. 3. Bounding box information detection during OCR

4   Performance Evaluation

In order to evaluate how accurately the OCR and ASR descriptions work, we use these descriptions for unsupervised clustering using K-Means clustering. For clustering, there are two types of validation statistics which can be used: internal validation and external validation. Internal validation uses internal information of the clustering process to evaluate the goodness of a clustering structure without reference to external information. These include metrics like Silhouette index, Dunn index, inertia etc. External cluster validation, which consists in comparing the results of a cluster analysis to an externally known result, such as externally provided class labels. It measures the extent to which cluster labels match externally supplied class labels. These include measures like purity, Rand index, Normalized Mutual Information etc. The metrics used in the current paper are inertia, Silhouette score and purity.

4.1   Evaluation Metrics

The evaluation metrics like Inertia, Silhouette Score and Purity are described here.

---

[6] https://pypi.org/project/Whoosh/.

Inertia. Intuitively, inertia tells how far away the points within a cluster are. Therefore, a small of inertia is aimed for. The range of inertia's value starts from zero and goes up.

Silhouette Score. The Silhouette Coefficient is calculated using the mean intra-cluster distance (a) and the mean nearest-cluster distance (b) for each sample. The Silhouette Coefficient for a sample is (b - a)/max (a, b). To clarify, b is the distance between a sample and the nearest cluster that the sample is not a part of. Note that Silhouette Coefficient is only defined if number of labels is greater than 2 and lesser than 1 less than the number of samples.

Purity. To compute purity, each cluster is assigned to the class which is most frequent in the cluster, and then the accuracy of this assignment is measured by counting the number of correctly assigned documents and dividing by N.

4.2   Result and Analysis

We perform two stages of experimentation in this section. In the first experiment, we use the raw frames and audio obtained from the video without any pre-processing like thresholding for frames and noise removal for audio. We compute the Inertia, Silhouette Score and Purity for clustering using only OCR generated text, ASR generated text and finally a combination of OCR and ASR generated text for clustering. The results are detailed in the Table 1. We find that the values of all the three metrics are generally lowest for OCR and highest for a combination of OCR and ASR. Thus, using a combination of OCR and ASR generated text to describe the video is a better alternative to using either of them alone.

The same set of experiments are performed with pre-processed frames and audio. The inertia, silhouette score and purity are calculated for OCR generated text, ASR generated text and combination of OCR and ASR. We observe that each of these scores are significantly better than the corresponding scores obtained using raw frames and audio. We also note that the best results are obtained using a combination of OCR and ASR generated text.

Table 1. Performance metrics for clustering using OCR, ASR and combination of OCR and ASR

	Without pre-processing			With pre-processing		
Metrics	**OCR**	**ASR**	**OCR+ASR**	**OCR**	**ASR**	**OCR+ASR**
Inertia	25.62	24.18	18.97	15.62	11.89	10.11
Silhouette score	0.64	0.69	0.75	0.76	0.82	0.84
Purity	0.69	0.71	0.73	0.78	0.82	0.83

5 Conclusion and Future Research Direction

In this paper, an attempt has been made to extract content from the videos using OCR and ASR. The extracted text is then used as a description for indexing and searching. In order to quantify how effectively the OCR and ASR generated texts can be used for indexing, we cluster the videos using K-Means which yields satisfactory results. The results indicate that a combination of OCR and ASR performs better than either of them individually.

The extraction of content can be made more robust if we incorporate object detection alongside OCR and ASR. Apart from text, frames also contain objects. These objects also encode information about the video. Hence, if we perform object detection on all the images, we can incorporate these objects into our embedding list. We can also optimise the results by differential weighting of the information obtained from OCR, ASR and Object Detection. The dataset can be improved by incorporating more genres and more videos to increase the scope of the dataset.

References

1. Medida, L.-H., Raman, K.: An optimized e-lecture video retrieval based on machine learning classification. Int. J. Eng. Adv. Technol. 8(6), 4820–4827 (2019)
2. Adcock, J., Cooper, M., Denoue, L., Pirsiavash, H., Rowe, L.: Talkminer: A Lecture Webcast Search Engine, pp. 241–250, October 2010
3. Balagopalan, A., Balasubramanian, L.L., Balasubramanian, V., Chandrasekharan, N., Damodar. A.: Automatic keyphrase extraction and segmentation of video lectures. In: 2012 IEEE International Conference on Technology Enhanced Education (ICTEE). IEEE, January 2012
4. Balasubramanian, V., Doraisamy, S.G., Kanakarajan, N.K.: A multimodal approach for extracting content descriptive metadata from lecture videos. J. Intell. Inf. Syst, 46(1), 121–145 (2015)
5. Chand, D., Ogul, H.: Content-based search in lecture video: a systematic literature review. In: 2020 3rd International Conference on Information and Computer Technologies (ICICT). IEEE, March 2020
6. Epshtein, B., Ofek, E., Wexler, Y.: Detecting text in natural scenes with stroke width transform. In: 2010 IEEE Computer Society Conference on Computer Vision and Pattern Recognition. IEEE, June 2010
7. Jeong, H.J., Kim, T.-E., Kim, M.H.: An accurate lecture video segmentation method by using sift and adaptive threshold. In: Proceedings of the 10th International Conference on Advances in Mobile Computing& Multimedia - MoMM 2012. ACM Press (2012)
8. Pranali, B., Anil, W., Kokhale, S.: Inhalt based video recuperation system using OCR and ASR technologies. In: 2015 International Conference on Computational Intelligence and Communication Networks (CICN). IEEE, December 2015
9. Yang, H., Meinel, C.: Content based lecture video retrieval using speech and video text information. IEEE Trans. Learn. Technol. 7(2), 142–154 (2014)
10. Yang, H., Quehl, B., Sack, H.: A framework for improved video text detection and recognition. Multim. Tools Appl. 69(1), 217–245 (2012). https://doi.org/10.1007/s11042-012-1250-6

GA-ProM: A Genetic Algorithm for Discovery of Complete Process Models from Unbalanced Logs

Sonia Deshmukh[1], Shikha Gupta[2(✉)], and Naveen Kumar[1]

[1] Department of Computer Science, University of Delhi, Delhi 110007, India
[2] Shaheed Sukhdev College of Business Studies, University of Delhi, Delhi 110007, India
shikhagupta@sscbsdu.ac.in

Abstract. Process mining is the practise of distilling a structured process description from a series of real executions. In past decades, different process discovery algorithms have been used to generate process models. In this paper, we propose a genetic mining algorithm (GA-ProM) for process discovery and compare it with other state-of-the-art algorithms, namely, α^{++}, genetic process mining, heuristic miner, and inductive logic programming. To evaluate the effectiveness of the proposed algorithm the experimentation was done on 21 synthetic event logs. The results show that the proposed algorithm outperforms the compared algorithms in generating the quality model.

Keywords: Process models · Genetic mining · Quality dimensions · Process mining · Event logs · Completeness

1 Introduction

Information systems record abundance of data to support automation of business processes in different domains, e.g., supply management, banking, software development, maintenance, and healthcare systems. A process is a collection of tasks or activities having one or more links between the activities. Actors perform the tasks to achieve the business goals. For example, actors in a banking system include bank manager, cashier, other employees, and the customers. The activities are performed by a subset of these actors to complete any process. Opening an account in any bank is a process where a customer performs the following activities—selecting the bank account type, choosing the right bank for creating account, gathering the required documents, completing the application process, collecting the account number, and depositing the funds into the account. Each organisation has detailed description of their processes. For example, in a banking system the documents required for opening an account, transaction limits are clearly described. In any process flow there may be areas of improvement that allow reduction in the overall working time/cost of the system, improve productivity and quality, balance resource utilization [18]. Thus arises the need

© Springer Nature Switzerland AG 2022
S. Sachdeva et al. (Eds.): BDA 2021, LNCS 13167, pp. 209–218, 2022.
https://doi.org/10.1007/978-3-030-96600-3_15

of a technique which could understand the flow and deviations in the processes. Process mining techniques facilitate discovery of process models from the data stored in the form of an event log and use these for further analysis. In the real world, process mining has found applications in the domains of healthcare, information and communication technology, manufacturing, education, finance, and logistics [18].

In this paper, we present a genetic mining approach for discovering process models. The experimentation was done on 21 unbalanced event logs and the proposed algorithm was compared with state-of-the-art algorithms, namely, Genetic process mining, Heuristic miner, α^{++} and Inductive logic programming. The results show that the proposed algorithm generates competitive process models as compared to the compared algorithms and it produces "good" quality process models for all the datasets.

The rest of the paper is organized as follows: Sect. 2 gives a brief summary of process mining concepts and the related work. Section 3 describes the proposed approach. Section 4 outlines the experimentation and explains the results. Section 5 concludes the paper.

2   Process Mining and the Related Work

Process mining algorithms abstract the event log data into a well-structured form known as a process model. An event log is a collection of cases that represents a process. A case, also known as a trace, is a set of events [16]. Each row of the event log represents an occurrence that pertains to a single case. Events are ordered within a case and have attributes such as case ID, activity/task name, and timestamp, among other things. Table 1 depicts an example event log with three tasks and three scenarios. Each distinct task is assigned an integer symbol—'Search product' (1), 'Check availability' (2), and 'Order product' (3). The timestamp indicates the occurrence time of the corresponding event. A trace is formed by picking all the events with the same Case ID and arranging them in ascending order of their timestamp. Each trace (which may include numerous tasks) can be any length and is expressed as a series of integer symbols. In the example log, the traces corresponding to the Case ID 7011 (events 1, 2, 5), 7012 (events 3, 7), and 7013 (events 4, 6) are 123, 12, and 13 respectively. Process mining approaches can be divided into three categories:

- Process discovery: the process of creating a process model from an event log without any prior knowledge [2,17].
- Conformance checking: determining whether the model provided by an algorithm matches the given event log [2,17].
- Enhancement: using information from the original process captured in an event log, improve or extend an existing process model [2,17].

In this paper, we focus on process discovery algorithms. Different state-of-the-art process discovery techniques include α [1], α^+ [15], Multi-phase miner [10,19], Heuristics miner [22], Genetic process mining (GPM) [16], α^{++} [23], $\alpha^{\#}$

Table 1. An example event log

Case ID	Activities/Tasks	Timestamp
7011	Search product	15-6-2020
7011	Check availability	16-6-2020
7012	Search product	15-6-2020
7013	Search product	15-6-2020
7011	Order product	17-6-2020
7013	Check availability	16-6-2020
7012	Order product	17-6-2020

[24], α^* [14], Fuzzy miner [3], Inductive logic programming (ILP) [12] etc. A few of these algorithms have explored the use of evolutionary methodology for discovering process models [11,16,21]. The recent algorithms in the domain of process discovery include Evolutionary tree miner (ETM) [6], Inductive miner [13], Multi-paradigm miner [8], a hybrid process mining approach that integrates the GPM, particle swarm optimization (PSO), and discrete differential evolution (DE) techniques to extract process models from event logs [7], Fodina algorithm which is an extension of Heuristic miner [4], Binary differential evolution [9]. Most of these algorithms either propose a hybrid technique using multiple evolutionary approach or optimize a weighted function of popular quality metrices of the process model.

3   Proposed Genetic Algorithm for Process Discovery (GA-ProM)

The process mining data is encapsulated in the form of an event log. Event log is the starting point for process discovery algorithms. An individual in the initial population is generated from the event log in the form of a causal relation matrix. Each bit of an individual is either 0 or 1. The steps for the proposed genetic mining algorithm for process discovery are outlined in the Algorithm 2. These steps are explained as follows:

The first task is to compute the dependency links between the tasks/activities in an event log V with n tasks/activities. The dependency relations that represent domain knowledge are stored in the form of the dependency measure matrix D [16]. Dependency measure matrix D which provides information about tasks in self loops, short loops (length-one and length-two loops) and tasks in parallel, is then used to generate causality relation matric M (Algorithm 1) [16]. Each individual in the initial population is represented by a causality relation matrix, which is computed as:

$$m_{t1,t2} = \begin{cases} 1 \text{ if } r < D(t1,\ t2,\ V)^P \\ 0, \text{ otherwise} \end{cases} \tag{1}$$

where $t1$, $t2 \in [1, n]$, $r \in [0, 1)$ is an n × n matrix of random numbers, generated for each individual in the population of N individuals. P regulates the influence of the dependency measure while creating the causality relation. [16] suggested the value of P = 1. A process mining algorithm should, in an ideal situation, generate a model that is simple (simplicity), precise (precision), general (generalization) and fits the available logs (completeness). All the quality dimensions are measured on the scale [0, 1]. Completeness [16] is a measure of the ability of the mined process model to reproduce the traces in the event log. Preciseness [20,21] is a metric for determining how much extra behaviour a model generates that isn't visible in the event log. Simplicity indicates the number of nodes in the mined model [20,21]. Generalization [11] measures the degree of which the mined model can reproduce future unforeseen behaviour.

Algorithm 1. Computation of Initial population

1: **for** i=1:N **do**
2:      M^i =zeros(n,n)
3:         r=rand(n,n), $r \in [0, 1)$
4:      **for** every tuple (t1, t2) in n x n do **do**
5:            $m^i_{t1,t2} \leftarrow (r < D(t1,t2,V)^P)$
6:      **end for**
7: **end for**

Algorithm 2. Pseudocode of the Proposed Genetic Algorithm for Process Mining (GA-ProM)

1: Generate N individuals (n x n dimension each) from the given event log
2: Evaluate each individual and find best solution
3: **while** Stopping criteria not met **do**
4:      **for** 1 to number of generations **do**
5:          Apply Binary Tournament Selection
6:          Create child population after applying Crossover and Mutation
7:          Repair the population
8:          Combine initial population and repaired population
9:          Apply Elitism to find fittest individuals
10:      **end for**
11: **end while**
12:   Return the best solution

In tournament selection, given a population of N individuals, a binary tournament amongst its members is used. Two individuals are randomly selected from the population and every individual must play two times. Out of those 2N randomly selected individuals (N pairs), the best individual is selected on the basis of quality dimensions under considerations. The winner individual goes into selection pool in the next iteration. Then, on the basis of a random crossover point, the causal relation of an individual is replaced with the casual relation of another randomly chosen individual at that crossover point. Input and output tasks are upgraded in causality matrix. The individuals are then mutated by adding a causality relation, deleting a causality relation or redistributing the relations in input or output sets of casual matrices. The mutations, that are inconsistent with the event log, are rejected. The proposed algorithm stops when there is no improvement in solutions for ten generations in a row, or it executes for the maximum number of specified generations.

4   Experimentation and Results

The population size is set to 100. The proposed algorithm (GA-ProM) is run for maximum of 100 iterations. The total number of runs is fixed at 30 and the results are based on the average performance of these runs. We have picked 21 complex case scenarios without noise [21]. The logs used in this experimentation are imbalanced, i.e., they contain traces with very different frequencies. The event logs were present in the .xes, .xml, and .hn formats. Using Prom 6 we first converted the .xes file into .csv format then traces are formed by picking all the events with the same Case ID and arranging them in ascending order of their timestamp to give the desired input file for experimentation. Summary of the experimental datasets is given in the Table 2.

The proposed algorithm (GA-ProM) is run on 21 unbalanced logs (Table 2), namely, ETM, g2–g10, g12–g15, g19–g25 [21]. These logs contain traces of varying frequencies. As a result, these logs are better for evaluating an algorithm's ability to overfit or underfit the data. The proposed algorithm is compared with α^{++} [23], Heuristic miner [22], Genetic miner [16], and ILP [12] algorithms. The values for completeness, preciseness, and simplicity for ETM, g2–g10, g12–g15, g19–g25 datasets are taken as reported by [21]. However, [21] do not report the value of generalization for these datasets. The value of generalization for ETM, g2–g10, g12–g15, g19–g25 datasets, for the models generated by α^{++}, Heuristic miner, Genetic miner, and ILP algorithms, is computed using the Cobefra tool [5,21]. The parameter settings for each of these algorithms is provided in Table 3.

4.1   Results and Analysis

In this paper, we compare the proposed algorithm (GA-ProM) with α^{++}, Heuristic miner, Genetic process mining (GPM), and ILP algorithms. α^{++} is an abstraction-based algorithm that deals with non-free choice constructs. It is an extension of one of the earliest process discovery algorithm, known as α-algorithm [23]. Heuristic Miner is heuristic-based algorithm and also an extension of α-algorithm. The frequency of activity in the traces is taken into account by heuristic miner to deal with noise and imperfection in the logs [22]. GPM, a main contender for the proposed algorithm, is a pure genetic approach proposed in process mining domain. GPM optimizes a function of completeness and

Table 2. Details of the datasets characteristics

Event-log name	Activities	Traces	Events	Source
ETM	7	100	790	[21]
g2	22	300	4501	
g3	29	300	14599	
g4	29	300	5975	
g5	20	300	6172	
g6	23	300	5419	
g7	29	300	14451	
g8	30	300	5133	
g9	26	300	5679	
g10	23	300	4117	
g12	26	300	4841	
g13	22	300	5007	
g14	24	300	11340	
g15	25	300	3978	
g19	23	300	4107	
g20	21	300	6193	
g21	22	300	3882	
g22	24	300	3095	
g23	25	300	9654	
g24	21	300	4130	
g25	20	300	6312	

Table 3. Parameter settings for the algorithms

Algorithm	Parameters
HM [21,22]	Relative-to-best–0.05, length-one-loops–0.9, length-two-loops–0.9, dependency–0.9, long distance–0.9
α^{++} [21,23]	–
ILP [12,21]	ILP Solver–(JavaILP, LPSolve 5.5), ILP Variant–Petri net, number of places–per causal dependency
GPM [16,21]	Selection–Binary tournament selection, Mutation Rate–0.2, Crossover Rate–0.8, Elitism Rate–0.2, extra behavior punishment–0.025
GA-ProM	Selection–Binary tournament selection, Mutation Rate–0.2, Crossover Rate–0.8

preciseness. ILP was created to add artificially generated negative events, such as the traces describing a path that is not permitted in a process [12]. A process model with "good" completeness value indicates that the process model is

Table 4. Quality dimensions for the process models obtained using α^{++}, HM, GPM, ILP, and the proposed algorithm (GA-ProM) for synthetic datasets (C: Completeness, P: Preciseness, S: Simplicity, G: Generalization)

Algorithm		ETM	g2	g3	g4	g5	g6	g7	g8	g9	g10	g12	g13	g14	g15	g19	g20	g21	g22	g23	g24	g25
Heuristic miner	C	0.37	1	1	0.78	1	0.66	1	0.52	0.74	0.78	1	1	0.91	0.87	0.85	1	1	0.9	0	0.93	0.23
	P	0.98	1	1	1	1	0.99	1	1	1	1	1	1	0.99	1	1	1	1	1	0.42	0.97	0.97
	S	1	1	1	1	1	0.99	0.98	0.93	0.96	1	1	1	1	1	1	1	1		0.93	1	0.94
	G	0.62	0.91	0.89	0.81	0.92	0.80	0.81	0.90	0.73	0.60	0.84	0.87	0.54	0.75	0.85	0.94	0.89	0.88	0.60	0.52	0.51
Alpha++	C	0.89	0.33	0	1	1	0.45	0	0.35	0.48	0.56	1	0.48	0	0.05	0.25	0.46	0.68	0.43	0	0	0.97
	P	1	0.96	0.18	0.97	1	1	0.12	1	1	1	0.97	1	0.82	0.14	0.98	0.86	0.09	1	0.42	0.11	0.26
	S	1	0.78	0.79	1	1	0.76	0.93	0.74	0.79	0.76	0.99	0.79	0.79	0.97	0.75	1	0.83	0.32	0.34	0.42	0.89
	G	0.56	0.62	0.74	0.91	0.92	0.84	0.81	0.91	0.59	0.43	0.94	0.95	0.61	0.72	0.72	0.97	0.89	0.65	0.36	0.33	0.45
ILP	C	1	1	1	1	1	1	1	1	1	1	1	1	1	1	1	1	1	1	1	1	1
	P	1	0.97	0.97	1	1	0.99	1	0.98	0.98	0.95	0.97	0.07	0.95	0.96	0.98	0.96	0.96	0.93	0.83	0.89	0.8
	S	0.93	0.93	0.92	0.96	1	0.74	0.93	0.67	0.9	0.68	0.99	0.97	0.69	0.76	0.79	0.79	0.97	0.52	0.28	0.7	0.59
	G	0.69	0.99	0.93	0.91	0.92	0.79	0.91	0.92	0.76	0.61	0.94	0.92	0.60	0.78	0.85	0.94	0.89	0.87	0.58	0.66	0.60
Genetic miner	C	0.3	1	0.31	0.59	1	1	1	0.26	0.48	0.48	1	0.75	1	0.15	0.2	1	1	0.43	0.2	0.72	0.41
	P	0.94	1	0.6	1	1	1	1	0.15	1	1	1	1	0.81	0.14	0.08	1	1	0.96	0.42	0.99	0.75
	S	1	1	1	0.97	1	1	1	0.72	0.96	0.88	1	0.95	0.99	0.88	0.95	1	1	0.88	1		0.82
	G	0.56	0.91	0.88	0.90	0.92	0.80	0.91	0.88	0.75	0.61	0.91	0.97	0.56	0.50	0.82	0.71	0.89	0.90	0.42	0.47	0.49
GA-ProM	C	1	1	0.997	0.99	1	0.98	0.99	0.99	0.98	0.96	1	1	0.99	0.99	0.99	1	1	0.99	0.87	0.99	0.88
	P	1	1	0.93	0.96	1	0.86	0.93	0.92	0.83	0.75	1	1	0.85	0.99	0.97	1	1	0.98	0.64	0.85	0.69
	S	1	1	1	1	1	1	1	1	1	1	1	1	0.99	0.99	0.99	1	1	0.98	0.99	0.99	0.99
	G	0.88	0.92	0.94	0.91	0.94	0.92	0.95	0.89	0.92	0.90	0.91	0.92	0.95	0.88	0.87	0.93	0.89	0.82	0.94	0.91	0.93

Table 5. Weighted average of the quality dimensions for synthetic datasets

Algorithm	ETM	g2	g3	g4	g5	g6	g7	g8	g9	g10	g12	g13	g14	g15	g19	g20	g21	g22	g23	g24	g25
HM	0.485	0.993	0.992	0.816	0.994	0.722	0.984	0.618	0.776	0.8	0.987	0.99	0.89	0.88	0.87	0.995	0.992	0.914	0.15	0.907	0.36
α^{++}	0.882	0.435	0.132	0.991	0.994	0.546	0.143	0.473	0.552	0.602	0.992	0.58	0.17	0.179	0.38	0.57	0.66	0.48	0.086	0.066	0.869
ILP	0.972	0.992	0.986	0.99	0.994	0.963	0.988	0.967	0.972	0.942	0.992	0.92	0.942	0.962	0.97	0.976	0.986	0.948	0.899	0.94	0.92
GPM	0.423	0.993	0.429	0.675	0.994	0.985	0.993	0.335	0.578	0.561	0.99	0.80	0.951	0.23	0.296	0.978	0.99	0.55	0.286	0.74	0.474
GA-ProM	0.991	0.994	0.981	0.988	0.995	0.971	0.979	0.976	0.962	0.939	0.993	0.994	0.976	0.982	0.979	0.995	0.992	0.975	0.867	0.973	0.871

able to reproduce the behavior expressed in the event log [6]. Completeness may be considered an important metric to measure the quality of process models. It's important was highlighted by the authors of [6] when they assigned a 10 times higher weight to completeness than the weight assigned to other quality dimensions.

The proposed algorithm optimizes completeness to discover the process models. In order to assess the quality of the discovered process models, we also compute the remaining three quality dimensions—namely, preciseness, simplicity, and generalization. We have also computed a weighted average of the quality dimensions, so as to rank the algorithms based on the weighted average and to compare the quality of the model generated by different algorithms [6].

Table 4 shows the values of the quality dimensions for the process models discovered by different algorithms. The proposed algorithm is able to achieve completeness values for the discovered process models that are better or at least as good as those achieved by the other algorithms. The results show that the discovered process models exhibit "good" values for the other quality dimensions also. Table 5 and Fig. 1 show that in terms of the weighted average, the process models discovered by the proposed algorithm are better in 14 datasets out of 21 in contrast to the other algorithms.

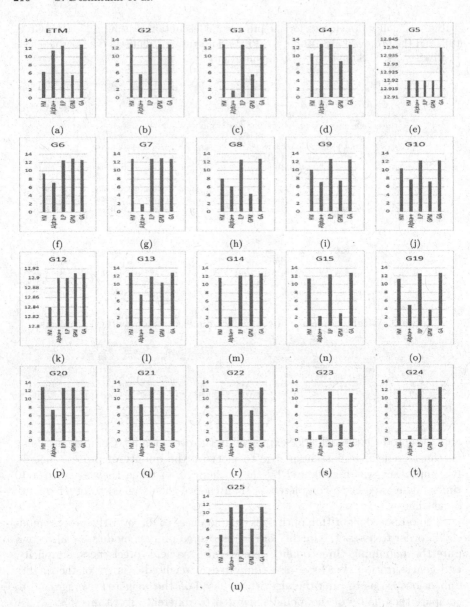

Fig. 1. Plots of weighted average of quality dimensions for the process models generated from Heuristic Miner, α^{++}, ILP, Genetic Process Miner, and the proposed algorithm (GA-ProM). In the figures GA-ProM has been abbreviated as GA.

5   Conclusion

In the past decade, the domain of process mining has developed into an active research area. Process mining techniques empower organizations to gain valuable insights into their business process from the digital data recorded in the informa-tion systems. The discovery of process models from the data stored in the form of an event log is usually the first step towards process improvements. The quality of the discovered process models is measured using four metrices, namely, simplicity, preciseness, generalization, and completeness. Completeness is considered more important than the others as a "good" completeness value indicates that the process model is able to reproduce the behavior expressed in the event log.

In this paper, we present a genetic mining algorithm (GA-ProM) for process model discovery from the given event data. We have experimented with completeness as the fitness function. The algorithm was tested on 21 unbalanced logs. We have also compared the proposed algorithm (GA-ProM) with the state-of-the-art algorithms, namely, Heuristic miner, α^{++}, Genetic process Mining (GPM), and Inductive logic programming (ILP). The results show the effectiveness of the proposed approach as it produces "good" quality process models for all the datasets. The proposed approach compares well with the other algorithms in terms of completeness value of the discovered process models and also in terms of the weighted average of the quality dimensions for the process models.

References

1. Van der Aalst, W., Weijters, T., Maruster, L.: Workflow mining: discovering process models from event logs. IEEE Trans. Knowl. Data Eng. **16**(9), 1128–1142 (2004)
2. van der Aalst, W.M.: Process Mining: Data Science in Action. Springer, Heidelberg (2016). https://doi.org/10.1007/978-3-662-49851-4
3. van der Aalst, W.M., Gunther, C.W.: Finding structure in unstructured processes: the case for process mining. In: Seventh International Conference on Application of Concurrency to System Design, ACSD 2007, pp. 3–12. IEEE (2007)
4. vanden Broucke, S.K., De Weerdt, J.: Fodina: a robust and flexible heuristic process discovery technique. Decis. Support Syst. **100**, 109–118 (2017)
5. vanden Broucke, S.K., De Weerdt, J., Vanthienen, J., Baesens, B.: A comprehensive benchmarking framework (CoBeFra) for conformance analysis between procedural process models and event logs in ProM. In: 2013 IEEE Symposium on Computational Intelligence and Data Mining (CIDM), pp. 254–261. IEEE (2013)
6. Buijs, J.C.A.M., van Dongen, B.F., van der Aalst, W.M.P.: On the role of fitness, precision, generalization and simplicity in process discovery. In: Meersman, R., Cruz, I.F. (eds.) OTM 2012. LNCS, vol. 7565, pp. 305–322. Springer, Heidelberg (2012). https://doi.org/10.1007/978-3-642-33606-5_19
7. Cheng, H.J., Ou-Yang, C., Juan, Y.C.: A hybrid approach to extract business process models with high fitness and precision. J. Ind. Prod. Eng. **32**(6), 351–359 (2015)
8. De Smedt, J., De Weerdt, J., Vanthienen, J.: Multi-paradigm process mining: retrieving better models by combining rules and sequences. In: Meersman, R., et al. (eds.) OTM 2014. LNCS, vol. 8841, pp. 446–453. Springer, Heidelberg (2014). https://doi.org/10.1007/978-3-662-45563-0_26

9. Deshmukh, S., Gupta, S., Varshney, S., Kumar, N.: A binary differential evolution approach to extract business process models. In: Tiwari, A., et al. (eds.) Soft Computing for Problem Solving. AISC, vol. 1392, pp. 279–290. Springer, Singapore (2021). https://doi.org/10.1007/978-981-16-2709-5_21

10. van Dongen, B.F., Van der Aalst, W.M.: Multi-phase process mining: aggregating instance graphs into EPCs and petri nets. In: PNCWB 2005 Workshop, pp. 35–58 (2005)

11. van Eck, M.: Alignment-based process model repair and its application to the Evolutionary Tree Miner. Ph.D. thesis, Master's thesis, Technische Universiteit Eindhoven (2013)

12. Goedertier, S., Martens, D., Vanthienen, J., Baesens, B.: Robust process discovery with artificial negative events. J. Mach. Learn. Res. **10**, 1305–1340 (2009)

13. Leemans, S.J.J., Fahland, D., van der Aalst, W.M.P.: Discovering block-structured process models from event logs - a constructive approach. In: Colom, J.-M., Desel, J. (eds.) PETRI NETS 2013. LNCS, vol. 7927, pp. 311–329. Springer, Heidelberg (2013). https://doi.org/10.1007/978-3-642-38697-8_17

14. Li, J., Liu, D., Yang, B.: Process mining: extending α-algorithm to mine duplicate tasks in process logs. In: Advances in Web and Network Technologies, and Information Management, pp. 396–407 (2007)

15. Alves de Medeiros, A., Van Dongen, B., Van Der Aalst, W., Weijters, A.: Process mining: extending the α-algorithm to mine short loops. Technical report, BETA Working Paper Series (2004)

16. Alves de Medeiros, A.K.: Genetic process mining. CIP-Data Library Technische Universiteit Eindhoven (2006, printed in)

17. van der Aalst, W.M.P.: Process Mining: Discovery, Conformance and Enhancement of Business Processes, vol. 8, p. 18. Springer, Heidelberg (2011). https://doi.org/10.1007/978-3-642-19345-3

18. dos Santos Garcia, C., et al.: Process mining techniques and applications-a systematic mapping study. Expert Syst. Appl. **133**, 260–295 (2019)

19. van Dongen, B.F., van der Aalst, W.M.P.: Multi-phase process mining: building instance graphs. In: Atzeni, P., Chu, W., Lu, H., Zhou, S., Ling, T.-W. (eds.) ER 2004. LNCS, vol. 3288, pp. 362–376. Springer, Heidelberg (2004). https://doi.org/10.1007/978-3-540-30464-7_29

20. Vázquez-Barreiros, B., Mucientes, M., Lama, M.: A genetic algorithm for process discovery guided by completeness, precision and simplicity. In: Sadiq, S., Soffer, P., Völzer, H. (eds.) BPM 2014. LNCS, vol. 8659, pp. 118–133. Springer, Cham (2014). https://doi.org/10.1007/978-3-319-10172-9_8

21. Vázquez-Barreiros, B., Mucientes, M., Lama, M.: ProDiGen: mining complete, precise and minimal structure process models with a genetic algorithm. Inf. Sci. **294**, 315–333 (2015)

22. Weijters, A., van Der Aalst, W.M., De Medeiros, A.A.: Process mining with the heuristics miner-algorithm. Technische Universiteit Eindhoven, Technical report. WP 166, pp. 1–34 (2006)

23. Wen, L., van der Aalst, W.M., Wang, J., Sun, J.: Mining process models with non-free-choice constructs. Data Min. Knowl. Disc. **15**(2), 145–180 (2007)

24. Wen, L., Wang, J., Sun, J.: Mining invisible tasks from event logs. In: Dong, G., Lin, X., Wang, W., Yang, Y., Yu, J.X. (eds.) APWeb/WAIM -2007. LNCS, vol. 4505, pp. 358–365. Springer, Heidelberg (2007). https://doi.org/10.1007/978-3-540-72524-4_38

Experimenting with Assamese Handwritten Character Recognition

Jaisal Singh[1] , Srinivasan Natesan[1] , Marcin Paprzycki[2]([⊠]) ,
and Maria Ganzha[2,3]

[1] Indian Institute of Technology, Guwahati, Assam, India
natesan@iitg.ac.in
[2] Systems Research Institute Polish Academy of Sciences, Warsaw, Poland
marcin.paprzycki@ibspan.waw.pl
[3] Warsaw University of Technology, Warsaw, Poland

Abstract. While Optical Character Recognition has become a popular tool in business and administration, its use for Assamese character recognition is still in early stages. Therefore, we have experimented with multiple neural network architectures, applied to this task, to establish general understanding of their performance. For the experiments, we have generated an Assamese script dataset, with over 27k images. To this dataset, LeNet5, AlexNet, ResNet50, InceptionV3, DenseNet201, and hybrid LSTM-CNN models have been applied. Next, geometric transformations that included shifting image horizontally and vertically, and rotating them, were applied, to augment the available dataset. The augmented dataset was used in experiments and accuracy of applied methods has been studied. Best accuracy, obtained on the test data, reached 96.83%.

Keywords: Character recognition · Assamese language · Machine learning · Convolutional Neural Networks · Data augmentation

1 Introduction

Optical character recognition (OCR) refers to the conversion of machine-printed, or handwritten, symbols into a machine-encoded text. Lot of work has been done concerning application of OCR to Western languages. Moreover, progress has been reported for handwritten Indian scripts, like Devanagari, Gurumukhi, Bengali, Tamil, Telugu and Malayalam. However, character recognition for the Assamese language, which is spoken by an estimated 25 million people, is still at a rather preliminary stage. The aim of this contribution is to follow-up results reported in [2]. This time (1) a larger dataset was prepared (27,177 in comparison to 26,707 images), (2) this dataset was further augmented by applying shift and rotate operations, and (3) additional classes of experiments have been completed (including two additional neural network architectures), to better understand the relationship between the data and performance of various (standard) models.

In this context, content of this contribution has been organized as follows. In Sect. 2, we start from brief summary of pertinent state-of-the-art. Interested

S. Sachdeva et al. (Eds.): BDA 2021, LNCS 13167, pp. 219–229, 2022.
https://doi.org/10.1007/978-3-030-96600-3_16

readers should consult [2] for more details. Section 3 describes generation of the dataset, and the preprocessing steps that were applied to it. Architectures of neural network models, used in reported experiments, and details of the experimental setup are provided in Sect. 4. Next, Sect. 5 summarizes obtained experimental results. Finally, Sect. 6 summarizes the main findings and outlines possible future work directions.

2 Related Work

Recognition of handwritten characters poses challenges, among others, due to inconsistent skew, alignment and individual variability. Furthermore, problems are aggravated by presence of distortions, *e.g.*, smears, (partially) faded characters, etc. These challenges manifest themselves, for instance, in digitization of Indian handwritten scripts (historical documents, in particular). Here, it is worthy noting that many Indian scripts have over 500 different symbols used in their text. This count is further enhanced by different vowel modifiers, which can be used with consonants, producing a threefold combinations of consonant-vowels (only certain vowels and consonants can be joined). Furthermore, scripts like Telugu, Gujarati, Devanagari and Bangla contain additional complex symbols. These complex symbols can also use vowel modifiers, resulting in further increase in the total count of possible symbols that can be encountered within the text that is being digitized. Somewhat easier to handle are Punjabi and Tamil scripts, due to relatively small number of possible symbols (about 70 and 150 respectively).

In this context, in [14], Chaudhuri and Pal summarize various attempts to character recognition. They pointed out that the approaches can be divided into feature-based and ones that are applied to the images directly.

Among the prominent examples of work, Sukhaswamy et al. [3] designed a Multiple Neural Network Associative Memory (MNNAM) architecture, for recognition of printed Telugu characters. In [4], Ajitha and Rao proposed a feature (extraction) based approach for recognition of Telugu characters. In [5], authors reported 96.7% accuracy for recognizing Odia characters, using Hopfield Neural Network, with zoning. For the Tamil characters, a Convolutional Neural Network with two convolutional layers, two fully connected layers, and with ReLu activation function, was applied in [6].

In [7], Indira et al. suggested the use of neural networks for recognition and classification of typed Hindi characters. The proposed technique achieved recognition rate of 76–95% for different samples. Reported results also showed that, as the number of samples used for training increases, the recognition accuracy also improves. In [8], an approach based on Deep Convolutional Neural Network has been proposed, for offline recognition of handwritten Gurumukhi characters. Using this approach, an accuracy of 74.66% was achieved. Here, no feature extraction was used.

Only a few attempts have been made towards Assamese Character recognition. Sarma et al. [9] attempted handwritten recognition with the help of template matching, the simplest method of character recognition. Since the performance of this method is very sensitive to random noises in the images, 80% accuracy was reported. Bania et al. [10] applied the zoning concept to compute diagonal features and the Grey Level Co-occurrence Matrix, to extract texture features. They also performed word recognition using feature extraction and text segmentation. The best reported accuracy, was at 90.34%. Finally, in [2] convolution neural networks have been applied to Asamese handwritten character recognition. There, the best reported accuracy was at 94.62%. Current contribution continues research found in this publication.

3   Dataset Used in the Experiments

Assamese script, is a syllabic alphabet, used in Assam and nearby regions for many languages, such as Sanskrit, Khasi, Jaintia, etc. The current form of the script has been observed in rock inscriptions dating from the 5th century A.D. The current script bears many similarities to the Bengali script.

There are few, publicly available (online), datasets of Assamese script images. Therefore, for the work reported in [2], we have created an Assamese handwritten dataset by obtaining samples from the faculty and staff of IIT Guwahati (Dataset1), and used the Tezpur University dataset (Dataset2). Moreover, for the purpose of this work, we have adapted dataset, created by S. Mandal (Dataset3). Let us now briefly describe these datasets.

Dataset1: A group of over 200 students, faculty and staff members of IIT Guwahati provided handwritten samples, each containing the 52 Assamese characters (11 vowels and 41 consonants). This group consisted of persons representing different age, education levels, and both genders, thus leading to a broad variety of writing styles. A total of 10,994 images belonged to this dataset.

Dataset2: This dataset was created with the help of the, Tezpur University created, Online Handwritten Assamese Character dataset [11] (available from the UCI Machine Learning repository [12]). The original dataset consists of 8235 Assamese characters, collected from 45 different writers. A combined 183 symbols, consisting of 11 vowels, 41 consonants and 131 juktakkhors (conjunct consonants) have been collected from each person. However, since juktakkhors were present only within this dataset (not in Dataset1), they have not been used in the work reported here. Moreover, since in the original dataset, characters are stored by capturing movement of the pen on a digitizing tablet, they have been turned into images, to match data collected for the Dataset1.

Dataset3: This dataset was created within the context of work of Mandal [13]. It contains both characters and word samples. For the same reasons as above, word samples have not been used in our work. This dataset is also similar to the Dataset2, as it stores information about movements of the pen on the tablet. Here, writers consisted of senior high school students from Guwahati, and adults belonging to 18–25 age group, with a minimum bachelor qualification. Again, pen movements were turned into 13342 images of vowels and consonants.

When the three datasets were combined, they contained a total of 26,707 images of consonants and vowels. It should be noted that, overall, the combined dataset was well balanced, with each character appearing almost the same number of times. Here, let us state that, while the dataset is not openly available online, any interested party can obtain it from Srinivasan Natesan, at the email listed above (list of authors of this paper).

3.1 Data Preprocessing

Next, the combined dataset was preprocessed, by applying the same steps that were reported in [2]. Specifically, the following operations have been executed (interested parties should consult [2] for more details).

- Images form Dataset1 have been *binarized* using Otsu binarization. Since images from Dataset2 and Dataset3 have been created (by repeating the recorded steps) as black-and white, they did not require binarization.
- All images have been *normalized* to minimize effects of individual handwriting styles. Specifically, they were cropped and rescaled to a standard 96×96 size image.
- Smoothing, based on 3-point moving average filter method, was applied to individual strokes for images originating from Dataset2 and Dataset3.

4   Neural Network Architectures

In the initial work, we have experimented with four neural network architectures: (1) LeNet5, (2) ResNet50, (3) InceptionV3, and (4) DenseNet201. Since they have been already described, and since their standard versions and implementations were used, readers are asked to consult [2] for more details. With the extended dataset that was prepared for this work, we have decided to experiment with additional neural network architectures, to build a more comprehensive image of relationship between (Assamese) handwritten image recognition and popular neural network architectures. Let us now, briefly, describe the additional architectures that have been applied.

4.1   Long Short Term Memory Networks

Long Short Term Memory (LSTM) networks are a class of Recurrent Neural Networks, that can handle not only single points of input, like images, but also sequences of data, like video. A standard LSTM unit comprises of a cell, an input gate, an output gate, and a forget gate. The cell can remember values over random time intervals and the three gates regulate the flow of information in the cell. LSTM models are known to perform well with predictions based on time series data. They also have the added benefit of not facing the problem of vanishing gradient. To apply LSTM to image recognition, it has been combined with Convolutional Neural Network, to form a LSTM-CNN hybrid (following ideas described in [15]).

4.2   AlexNet

CNNs have become the preferred choice for image recognition. However, as the size of datasets began to increase, there arose a need to optimize the training to cut down the training time. Here, AlexNet, comprising of 8 layers: 5 convolutional layers, and 3 fully-connected layers, solved this problem and won the 2012 ImageNet competition. What makes AlexNet different from its predecessors is:

- Nonlinearity – *relu* was used as an activation function, instead of *tanh*, leading to reduction in the training time (by a factor of six).
- Mulitple GPUs – AlexNet trained the model simultaneously on two GPUs, thus significantly reducing train times. This is particularly important when bigger models are to be trained.
- Overlapping pooling – that resulted in reduction of the size of the network.

Finally, AlexNet uses Data augmentation and Dropout layers to reduce the problem of overfitting and to increase overall model performance.

4.3   Experimental Setup

The models were implemented using Python frameworks Keras and Tensorflow. In all experiments data was split into 80% for training and 20% for testing. Each result reported here is an average of 5 runs. Unless explicitly stated, experiments have been performed for different images drawn (randomly) to the training and the testing datasets.

In all experiments, training rate 0.001 and batch size 64 were used. This point deserves an explanation. While it is a well-known fact that CNN models can go through the process of hyper-parameter tuning, and this was done to some extent in the work reported in [2], this avenue was not pursued here. The reason is that, with additional data formats (shifted and rotated) the existing solution space for model tuning becomes very large. Hence we have decided to fix the learning rate and the batch size (at reasonable values) and to observe the "general behavior" of the six models under investigation. Further hyper-parameter tuning is one of directions of possible future research. However, it may not be the most fruitful and interesting, as we argue in Sect. 6.

5   Experimental Results and Analysis

5.1   Complete Original Dataset Results

The first set of experiments has been conducted by applying, above described, six neural network architectures to the "complete original dataset", consisting of Dataset1 + Dataset2 + Dataset3. Obtained results are summarized in Table 1. In the Table we represent: testing accuracy and information after how many epochs the training process was stopped, due to the lack of further progress. Moreover, to simplify the comparison, we copy the results from [2] (Table 1), available for the four architectures that were considered there. Here, we have to recognize that some results reported in [2] were obtained for different learning rates than these reported here (0.0008 for LeNet and ResNet). Moreover, all results have been obtained for different batch sizes (128 for ResNet, and 48 for the remaining three architectures). However, as stated above, our goal was to obtain the general idea on accuracy of different architectures, and to establish the "general baseline" as to what performance can be expected from application of modern neural networks directly to the problem (without feature extraction).

Table 1. Accuracy of models applied to the complete original dataset

Model	Test accuracy	Stop epochs	Earlier results
LSTM-CNN	74.62	100	
LeNet05	85.79	70	86.25
AlexNet	**93.04**	60	
InceptionV3	92.18	50	94.09
ResNet50	92.82	70	93.55
DenseNet201	**93.08**	50	**94.26**

A number of observations can be made. First, the performance of the LSTM-CNN hybrid is a clear outlier. This model does not work for the considered task (even after substantially longer training – 100 epochs). Similarly to the results reported in [2], LeNet5 is also not competitive (its accuracy is approximately 7% worse than that of the remaining models).

The performance of the four "competitive" models is very close to each other. The difference between AlexNet, Inception, ResNet and DenseNet is less than 1%. Since they were all trained using the same learning rate and batch size, it can be conjectured that after individual hyper-parameter tuning it is possible that their performance could be improved.

Finally, obtained results are slightly worse than those reported in [2]. The difference is also of order of 1% for all models. However, it should be kept in mind that the "old" results were obtained after some hyper-parameter tuning (and on a different, slightly smaller, dataset).

Overall, somewhat contradictory to the naive expectations, extending the size of the dataset did not result in immediate performance improvement. Therefore, we have decided to further explore how the dataset can be augmented to make the models generalize (and perform) better.

5.2   Dataset with Shifted Images

In the second phase of our work we have decided to augmented the existing dataset with images shifted in eight directions: Top, Bottom, Left, Right, Top

Left, Top Right, Bottom Left and Bottom Right, by 2 pixels. This resulted in creation of eight new datasets, with a total of 8 × 27,177 = 190,239 images. These image sets have been mixed with the original ones (each direction of shift separately) and then randomly split into 80-20 datasets for training and testing. Due to the extremely poor performance for the original dataset, in this round of experiments, we have skipped the LSTM-CNN model and experimented only with the remaining five architectures. The same learning rate (0.001) and batch size (64) were used. Taking advantage of the fact that the models have been already trained on the original dataset, we used standard transfer learning to reduce the training time for the shifted images. Obtained results are summarized in Table 2. Each result is an average of 5 runs for different data splits.

Table 2. Accuracy for shifted datasets

Model	Shift left	Shift right	Shift top	Shift bottom	Top right	Top left	Bottom right	Bottom left
LeNet05	79.22	79.31	79.68	78.41	80.89	79.97	79.71	79.88
Inception V3	94.44	94.58	94.65	94.55	94.69	94.85	94.71	94.77
ResNet 50	95.53	95.45	95.52	95.57	95.75	95.61	95.65	95.57
AlexNet	**95.85**	**95.78**	**95.88**	**95.81**	**96.18**	**95.99**	**95.98**	95.91
DenseNet201	95.52	95.73	**95.88**	**95.81**	95.98	95.91	95.94	**95.95**

Comparing results presented in Table 1 and Table 2 it is easy to observe that LeNet remains the least accurate of the 5 models. In all remaining cases, augmenting dataset by shifted images results in overall performance improvement. In the Table, we have marked, in bold, the best results for each shift-direction augmentation. This brings an interesting observation. AlexNet, which was one of "overall winners", reported in Table 1, is outperformed by ResNet and/or DenseNet.

Moreover, out of the four better performing models, for all directions of image shift, the performance difference remains of the order of 1%. Combining this with the fact that, in this series of experiments, AlexNet was outperformed by ResNet and/or DenseNet, it becomes even clearer that further hyper-parameter tuning may result in change of order of best performers. However, it can be also stated that these *four models are very much comparable in their performance.*

5.3   Dataset with Rotated Images

Seeing that augmentation of data with shifted images helped, we have tried use augmentation with images that are rotated by one and by two degrees in both clockwise and counterclockwise direction. This brought four new datasets with a total of 4× 27,177= 108,708 images. Rotated images have been (separately) mixed with the original dataset and randomly split into 80-20 ratio for training and testing. Again, for each dataset, 5 experiments have been run, applying transfer learning to reduce training time. Average accuracy is reported in Table 3.

Table 3. Accuracy for rotated datasets

Model	Clockwise 1 degree	Counter 1 degrees	Clockwise 2 degrees	Counter 2 degrees
LeNet05	78.44	77.78	78.47	78.64
Inception V3	94.24	94.19	94.55	94.61
ResNet 50	94.25	94.41	94.67	94.77
AlexNet	95.18	95.22	**96.01**	**96.22**
DenseNet201	**95.24**	**95.28**	95.75	95.85

Here, the general pattern of results, reported above, repeats, though with some differences. Again, LeNet is not competitive. Hence, it was not be considered in what follows. Among the remaining four models. ResNet and DenseNet are the "winners". However, the performance difference between all four models remains of order 1%. Interestingly, DenseNet outperforms ResNet for datasets with images rotated by one degree, while the reverse can be observed for images rotated by two degrees. We do not have an explanation for this behavior.

5.4   Experiments with Different Train and Test Datasets

In this set of experiments we have studied interplay between various data subsets, used for training and testing. All dataset categories, listed next, start with the dataset created by combining Dataset1 + Dataset2 + Dataset3. This *raw* dataset, consists of original images without normalisation, or smoothing (where it would have been applicable), so it includes all extra empty spaces around the symbols and rough images resulting from drawing letters on the basis of pen movements, i.e. images are exactly the way they were "produced" (before any preprocessing was applied). The *normalized* dataset consists of *raw* images that have been normalized (see, Sect. 3.1). The *no smoothing* dataset consists of images that have been normalized, but not smoothed. The *processed* dataset refers to the dataset that has been fully prepossessed, i.e. normalised and smoothed. The *shifted* dataset consists of all eight sets, in which images have been shifted. The *rotated* dataset includes four sets of images that have been rotated. In Table 4 presented are results obtained when treating separately *raw, normalized, no smoothing* and *processed* datasets, for the four remaining models. As previously, the same learning rate, batch size and 80-20 split were applied and each result is an average of 5 runs.

As can be seen, when raw dataset is used for both training and testing, performance of all models is lacking. Interestingly, but somewhat expectedly, when models are trained on raw dataset, but applied to processed dataset (for

Table 4. Accuracy for basic separate datasets

Train dataset	Test dataset	AlexNet	Inception	ResNet	DenseNet
Raw	Raw	53.82	55.84	55.18	57.92
Raw	Processed	68.82	66.42	64.88	66.71
Processed	Raw	62.64	61.49	62.79	65.18
Processed	No smoothing	94.08	93.58	93.58	94.25

testing), the performance improves by 10–15%. Here the ability to "capture" features of the "more general" dataset pays dividends when the model is applied to the less general (made more uniform by preprocessing) one. This can be seen further when processed dataset is used for training and applied to the raw dataset. This reduces the performance by 2–5%, depending on the model. Finally, the last line in the Table shows that smoothing has much weaker effect on the performance than normalization. Obviously, this can be related to the fact that not all images had to be smoothed (see, Sect. 3.1). When processed dataset is used for training and applied to the dataset that was normalized but not smoothed, the performance is similar to the one reported in Table 1. Here, as in all cases reported thus far, the performance of the four models is very close to each other. Hence, again, it can be expected that individual hyper-parameter tuning may influence the fact that AlexNet and DenseNet slightly outperformed the remaining two models.

The final group of experiments involved processed, shifted and rotated datasets, mixed with each other in various combinations and used for training and testing (as denoted in Table 5). The remaining aspects of the experimental setup have not been changed. Obtained results are summarized in Table 5.

Table 5. Accuracy for combined datasets

Train dataset	Test dataset	AlexNet	Inception	ResNet	DenseNet
processed+shifted	processed	95.49	94.93	94.79	95.55
processed	processed+shifted	93.78	93.04	93.03	93.79
processed+rotated	processed	95.98	95.18	95.28	96.08
processed	processed+rotated	93.58	94.78	93.09	93.15
processed +shifted+rotated	processed	**96.78**	95.35	95.68	96.71
processed +shifted+rotated	processed +shifted+rotated	96.47	**95.78**	**95.87**	*96.83*

As can be expected, the best results have been obtained for the largest datasets, consisting of all individual datasets combined together and then split 80-20 for training and testing (depicted in the last line in the Table). The only exception is AlexNet, which obtained its best performance when processed +' shifted + rotated data was used for training and applied to data that was processed (without the remaining two datasets included). Overall, the best performance was obtained by the DenseNet (which reached 96.83% accuracy). However, again, all results obtained by the four considered models are within 1% from each other, for each dataset that was experimented with (i.e. for each line in the Table).

6   Concluding Remarks

The aim of this work was to experimentally explore performance of neural network architectures applied directly (without feature extraction) to the recognition of Assamese characters (vowels and consonants). The main lessons learned can be summarized as follows: (1) LSTM-CNN hybrid and LeNet models are not competitive for this task; (2) without hyper-parameter tuning, using generic learning rate and batch size, the remaining four CNN's (AlexNet, Inception, DenseNet and ResNet) perform very closely to each other (for all experiments that performance difference was of order of 1%); (3) augmenting the dataset by shifted and/or rotated images (and thus increasing the size of the dataset) has definite positive effect on performance; (4) conservative baseline for accuracy of image recognition, without feature extraction, can be placed at around 95%.

Obviously, further work can be pursued including, among others, the following directions: (a) tuning hyper-parameters for each of the four best models, (b) building meta-classifiers, or (c) further augmenting the training dataset, possibly by applying GANNs to generate synthetic datasets.

However, above presented data brings about a more general reflection. When machine learning is applied to "well prepared", large datasets, performance of 95% or more has been reported. However, as can be seen in Table 4, when raw images are used in training and testing, the performance drops to about 57%. This means that, from practical point of vies, this "performance" is almost useless. In other words, if a similar method was to be applied to digitization of historical manuscripts, about 40% of characters would be misrepresented.

This also brings the question of recognition words within a text, which is an even more more complex endeavour. Specifically, this would involve pipelines, where words may or may not be split into individual characters and words/characters would have to be automatically preprocessed. Next, depending on the selected approach, an appropriate classifier would have to be trained. This can be relatively easily completed in the case of structured documents (see, for instance [1]), but is much more complex for unstructured text.

Overall, it can be stated that, while the recognition of prepossessed scripts representing symbols in majority of languages is already well understood, the main research directions concern (i) recognition of characters represented as raw images, and (ii) recognition of words originating form unstructured handwritten documents.

References

1. Denisiuk, A., Ganzha, M., Paprzycki, M., Wasielewska-Michniewska, K.: Feature extraction for polish language named entities recognition in intelligent office assistant. In: Proceedings of the 55th Hawaii International Conference on System Sciences (in press)
2. Mangal, D., Yadav, M., Natesan, S., Paprzycki, M., Ganzha, M.: Assamese character recognition using convolutional neural networks. In: Proceedings of 2nd International Conference on Artificial Intelligence: Advances and Applications, Algorithms for Intelligent Systems (2022). https://doi.org/10.1007/978-981-16-6332-1_70
3. Sukhaswamy, M.B., Seetharamulu, P., Pujari, A.K.: Recognition of telugu characters using neural network. Int. J. Neural Syst. 6(3), 317–357 (1995)
4. Rao, P.V.S., Ajitha, T.M.: Â Telugu script recognition - a feature based approach. In: Proceedings of Third International Conference on DAR vol. 1 (1995)
5. Jena, O.P., Pradhan, S.K., Biswal, P.K., Tripathy, A.K.: Odia characters and numerals recognition using hopfield neural network based on zoning feature. Int. J. Recent Technol. Eng. 8, 4928–4937(2019)
6. Prakash, A.A., Preethi, S.: Isolated offline tamil handwritten character recognition using deep convolutional neural network. In: International Conference on Intelligent Computing and Communication for Smart World (I2C2SW), Erode, India, pp. 278–281 (2018)
7. Indira, B., Shalini, M., Ramana Muthy, M.V., Shaik, M.S.: Classification and recognition of printed Hindi characters using ANN. Int. J. Image Graph. Signal Process. 6, 15–21 (2012)
8. Jindal, U., Gupta, S., Jain, V., Paprzycki, M.: Offline handwritten Gurumukhi character recognition system using deep learning. In: Jain, L.C., Virvou, M., Piuri, V., Balas, V.E. (eds.) Advances in Bioinformatics, Multimedia, and Electronics Circuits and Signals. AISC, vol. 1064, pp. 121–133. Springer, Singapore (2020). https://doi.org/10.1007/978-981-15-0339-9_11
9. Sarma, P., Chourasia, C.K., Barman, M.: Handwritten assamese character recognition. In: IEEE 5th International Conference for Convergence in Technology (I2CT), pp. 1–6. Bombay, India (2019)
10. Bania, R.K., Khan, R.: Handwritten Assamese character recognition using texture and diagonal orientation features with artificial neural network. Int. J. Appl. Eng. Res. 13(10), 7797–7805 (2018)
11. Baruah, U., Hazarika, S.M.: A Dataset of Online Handwritten Assamese Characters. J. Inf. Process. Syst. 11(3), 325-341 (2015). https://doi.org/10.3745/JIPS.02.0008
12. Dua, D., Graff, C.: UCI Machine Learning Repository, University of California, School of Information and Computer Science, Irvine (2019). http://archive.ics.uci.edu/ml
13. Mandal, S.: Noval approaches for basic unit modeling in online handwritting recognition. PhD Thesis, Department of Electronics and Electrical Engineering, Indian Institute of Technology (2019)
14. Pal, U., Chaudhuri, B.B.: Indian script character recognition: a survey. Patt Recogn. 37(9), 1887–1899 (2004). https://doi.org/10.1016/j.patcog.2004.02.003
15. Zhang, J., Li, Y., Tian, J., Li, T.: LSTM-CNN hybrid model for text classification. In: 2018 IEEE 3rd Advanced Information Technology, Electronic and Automation Control Conference (IAEAC), pp. 1675–1680 (2018). https://doi.org/10.1109/IAEAC.2018.8577620

Business Analytics

Approximate Fault-Tolerant Data Stream Aggregation for Edge Computing

Daiki Takao, Kento Sugiura, and Yoshiharu Ishikawa$^{(\boxtimes)}$

Graduate School of Informatics, Nagoya University,
Furo-cho, Chikusa-ku, Nagoya 464-8601, Japan
takao@db.is.i.nagoya-u.ac.jp, {sugiura,ishikawa}@i.nagoya-u.ac.jp

Abstract. With the development of IoT, edge computing has been attracting attention in recent years. In edge computing, simple data processing, such as aggregation and filtering, can be performed at network edges to reduce the amount of data communication and distribute the processing load. In edge computing applications, it is important to guarantee low latency, high reliability, and fault tolerance. We are working on the solution of this problem in the context of environmental sensing applications. In this paper, we outline our approach. In the proposed method, the aggregate value of each device is calculated approximately and the fault tolerance is also guaranteed approximately even when the input data is missing due to sensor device failure or communication failure. In addition, the proposed method reduces the delay by outputting the processing result when the error guarantee satisfies the user's requirement.

Keywords: Stream processing · Edge computing · Approximate processing · Fault tolerance

1 Introduction

With the spread of *Internet of Things* (*IoT*), collection and aggregation of information via networks is becoming increasingly important. The approach of transferring a large amount of information obtained by sensing to a server and processing it all at once by the server has a large communication overhead and greatly deteriorates the processing speed. Therefore, *edge computing*, which performs filtering and aggregation at the edge of the network as shown in Fig. 1, is attracting much attention to solve this problem.

Distributed stream processing that introduces processing at edges has advantages in terms of efficiency, but it also has difficulties. In particular, *fault-tolerance* is an important issue, especially how to process when a network failure occurs. One solution is to multiplex resources, but in edge environments, unlike cluster environments where resources are plentiful, sensing data management and stream processing must be performed by a small number of poorly equipped machines, and fault-tolerance based on the premise of resource saving is required.

© Springer Nature Switzerland AG 2022
S. Sachdeva et al. (Eds.): BDA 2021, LNCS 13167, pp. 233–244, 2022.
https://doi.org/10.1007/978-3-030-96600-3_17

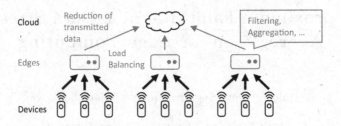

Fig. 1. Overview of edge computing

In this study, we focus on *approximate stream processing*. The basic idea is to reduce the resource burden of stream processing by allowing a small amount of error, instead of the considerable overhead incurred when trying to do it rigorously. Since the information obtained from sensors and other devices contains errors by nature, a certain amount of error is considered acceptable in many cases. Approximate stream processing has been the subject of much theoretical and systematic research, and approaches to processing with error guarantees have been proposed [5,9]. In our research, we aim to develop these ideas and establish a method that is superior in terms of quality and efficiency.

In this paper, we outline our quality-assured approximate stream processing approach for environmental sensing applications such as temperature. In environmental sensing, the acquired values of each sensor are often correlated, and efficiency improvement using correlation between sensors has been proposed in the context of sensor networks [4]. We extend that approach by approximating the aggregate value of each sensor to provide a theoretical error guarantee for approximate processing results, which can be applied to fault-tolerance and efficient processing.

The conceptual of the proposed method is shown in Fig. 2. The system receives the required confidence and tolerance values from the user and estimates the aggregate values within the range that satisfies the user's requirements, and reduce the processing delay. As an example, we consider a sensor stream with missing data as shown in Fig. 3. Note that in this example, the data from device X_2 has not reached the edge since time 4. Therefore, time 3 is the latest point in time (*watermark*) for the entire data source, and data processing cannot proceed after time 4. Therefore, approximate aggregate values are calculated by estimating the data of device X_2 after time 4 based on the statistical model that models the correlation and the measured values of other devices. In addition, the loss of internal state due to node failure can be regarded as data loss of all devices, such as the situation at time 5 and 6. Therefore, by estimating the missing values using a statistical model, we can approximate and guarantee the fault-tolerance of stream processing.

2   Preliminaries

In this section, we define the terms and concepts needed for the rest of this paper.

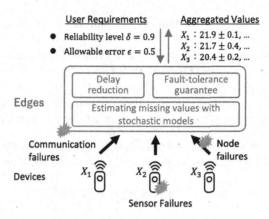

Fig. 2. Overview of the proposed method

device	time step							
	1	2	3	4	5	6	7	⋯
X_1	21	22	22	22	-	-	23	
X_2	21	21	22	-	-	-	-	⋯
X_3	-	20	20	20	-	-	21	

Fig. 3. Data stream with missing items

2.1 Definition of Data Streams

First, we define a *sensor data stream*.

Definition 1. *Let $X = \{X_1, X_2, \ldots, X_n\}$ be the set of all n devices. All devices are assumed to output data periodically and synchronously, and the true values of devices at each time $t \in \mathbb{N}^+$ is denoted by $x^t = \langle x_1^t, x_2^t, \ldots, x_n^t \rangle$. he sensor data stream that started to be measured at a certain time t is represented as an infinite series of true values of the devices at each time by the following formula:*

$$x^{[t,\infty)} = \langle x^t, x^{t+1}, \ldots \rangle \tag{1}$$

However, in an edge computing environment, data from all devices may not be available at each time. Therefore, we define a *data stream with missing items* as a data stream that is actually observed.

Definition 2. *Let the observed values at time t be $o^t = \langle o_1^t, o_2^t, \ldots, o_n^t \rangle$. For a data stream $x^{[t,\infty)}$, we define a data stream with missing items by the following formula:*

$$o^{[t,\infty)} = \langle o^t, o^{t+1}, \ldots \rangle, \tag{2}$$

where it is assumed that the observed values are always equal to the true values.

In this paper, we assume device failures, communication failures, and node failures as missing scenarios.

Figure 3 shows a data stream with missing items. For example, at time 1, the value of device X_3 is missing. On the other hand, since the values of X_1 and X_2 are observed, $o_1^1 = x_1^1 = 21$ and $o_2^1 = x_2^1 = 21$ hold.

Next, we define partitioning of a data stream based on *sliding windows*, assuming an aggregation process with a certain time window length.

Definition 3. *Given a data stream $x^{[t,\infty)}$, a window width w, and a sliding width l, using the sliding window concept, the data stream is partitioned based on the following formula:*

$$\left\{ x^{[t',t'+w)} \mid \forall i \in \mathbb{N}^0, t' = t + i \cdot l \right\}. \tag{3}$$

For example, if we parttion the data stream in Fig. 3 with the window width $w = 3$ and sliding width $l = 2$, we obtain the following set:

$$\left\{ \langle o^1, o^2, o^3 \rangle, \langle o^3, o^4, o^5 \rangle, \dots \right\}. \tag{4}$$

In this paper, we focus on the processing of a observed finite stream $o^{[t,t+w)}$ partitioned into time windows.

2.2 Estimation of Missing Values Based on Multidimensional Gaussian Distribution

In this research, the correlation between devices X is represented using a *multidimensional (multivariate) Gaussian distribution $N(x \mid \mu, \Sigma)$*. Note that μ and Σ represent the mean vector and variance-covariance matrix of X, respectively. In the following, the missing value estimation by range integration and the improvement of the estimation accuracy by using the posterior probability distribution (correlation between devices) are explained.

When the value of the device $X \in X$ follows a Gaussian distribution with expected value μ and variance σ, the true value can be estimated probabilistically by range integration even if the value of X is missing. For example, the probability that the value of X is x' can be obtained using a certain error upper bound e as follows:

$$P(X \in [x' - e, x' + e]) = \int_{x'-e}^{x'+e} N(x \mid \mu_X, \sigma_X) dx. \tag{5}$$

When using the same error upper bound, since the probability is largest when $x' = \mu$, we can simply use the expected value as an estimate to recover the missing value probabilistically.

There is a correlation between devices in the real world, for example, the measured values of temperature sensors located geographically close to each other have similar values. In [4], this correlation is modeled by a multidimensional Gaussian distribution, and a method to improve the estimation accuracy by using the posterior probability distribution has been proposed. When the observed value o is obtained for some device $O \subseteq X$, the Gaussian distribution $\mathcal{N}(x \mid \mu_{X|o}, \sigma_{Xo}, \sigma_{X|o})$ can be obtained from the following equation [3]:

$$\mu_{X|o} = \mu_X + \Sigma_{XO}\Sigma_{OO}^{-1}(o - \mu_O) \tag{6}$$

$$\sigma_{X|o} = \Sigma_{XX} - \Sigma_{XO}\Sigma_{OO}^{-1}\Sigma_{OX}, \tag{7}$$

where the subscript of each symbol indicates that the dimension corresponding to the subscript has been extracted from the vector or matrix. For example, μ_X is the value of the dimension corresponding to X extracted from the mean vector μ (i.e., the expected value of device X), and Σ_{XO} is the columns corresponding to O extracted from the X rows of the variance-covariance matrix Σ (i.e., the covariance vector of device X and the set of devices O from which the data was obtained). Since the variance of the posterior probability distribution is reduced compared to that of the prior distribution, more accurate estimation is possible.

3 Estimation of Time Window Aggregation Based on Probabilistic Models

In [4], Deshpande et al. proposed an approximate aggregate query method using a multidimensional Gaussian distribution. The method targets the aggregation among all devices X, i.e., the processing for a specific time. However, stream processing in this paper generally involves aggregation over time windows, which requires processing over multiple time periods. We extend the method of Deshpande et al. to aggregation over time windows and derive the estimated confidence of the aggregation process for data streams with missing items [12]. Since the proposed method uses the reproducibility of the Gaussian distribution, it covers both mean and total aggregates. However, due to space limitations, only the mean aggregate will be presented below.

3.1 Derivation of Gaussian Distribution for Aggregate Values

Since the Gaussian distribution is reproducible, the mean value of the device $X \in X$ in a given time window $x^{[t,t+w)}$, $Y_X = (\sum_{t' \in [t,t+w]} X^{t'})/w$, also follows a Gaussian distribution. By replacing the expression for the expected value and variance of the Gaussian distribution for the aggregate value at a particular time in [4] with Y_X, we can derive the expected value and variance of Y_X as follows [12]:

$$E[Y_X] = \frac{1}{w} \sum_{t' \in [t,t+w)} E[X^{t'}] \tag{8}$$

$$E[(Y_X - \mu_{Y_X})^2] \approx \frac{1}{w^2} \sum_{t' \in [t,t+w)} E\left[(X^{t'} - \mu_{X^{t'}})^2\right]. \tag{9}$$

In the approximate derivation of Eq. (9), it is assumed that the same probability model $\mathcal{N}(\mu, \Sigma)$ is used for all times in the time window.

As shown in Eq. (8) and (9), the average aggregate for the time window $x^{[t,t+w)}$ is obtained from the expectation and variance of the device at each time. Therefore, from the partial observations o^t obtained at each time, the expectation and variance of the missing values are calculated based on Eq. (6) and (7), and the mean value for the entire time window is estimated. If there is an observed value o^t at time t for the device $X \in \boldsymbol{X}$, we calculate the expected value as $\mu_X^t = o^t$ and the variance as $\sigma_X^t = 0$.

3.2   Confidence of Estimation in Time Window Aggregation

Since the results of the time window aggregation also follow a Gaussian distribution as described in the previous section, we can calculate the confidence of the estimation in a similar way to Sect. 2.2. The estimated confidence of the mean value Y_X of the device $X \in \boldsymbol{X}$ in a given time window is given by the following equation:

$$P(Y_X \in [\mu_{Y_X} - e, \mu_{Y_X} + e]) = \int_{\mu_{Y_X} - e}^{\mu_{Y_X} + e} \mathcal{N}(y \mid \mu_{Y_X}, \sigma_{Y_X}) dy, \tag{10}$$

where the expected value μ_{Y_X} is treated as an estimate, and the upper bound on the error is denoted by e.

Note here that the results of the aggregate query are probabilistic because they contain missing values. Therefore, in order to calculate the estimated confidence level, the user needs to specify the parameters for either the error upper bound or the required confidence level. For example, suppose that the user specifies the required confidence level δ as a parameter. In this case, by solving $P(Y_X \in [\mu_{Y_X} - e, \mu_{Y_X} + e]) = \delta$, we can calculate the minimum estimation error e that satisfies the required confidence level. Note that if the number of missing values is extremely large, the estimation error e also becomes large, and the range of possible values of the aggregate $\mu_{Y_X} \pm e$ also becomes wide.

4   Delay Reduction Based on Probabilistic Estimation

The approach of approximate data stream processing can be applied to efficient processing as well as fault handling. In this section, we discuss delay reduction in time window aggregation based on the output of estimates. The idea here is to reduce the delay in the time window aggregation by outputting the results when the user-specified estimation accuracy is met.

In this study, we use the *required confidence level* δ and *acceptable error* ϵ as thresholds to determine the estimation accuracy. In other words, we require that the aggregate value of each device in a given time window satisfies the following formula:

$$\forall X \in \boldsymbol{X}, P(Y_X \in [\mu_{Y_X} - \epsilon, \mu_{Y_X} + \epsilon]) \geq \delta. \tag{11}$$

Note, however, that this equation may not be satisfied depending on the degree of deficiency, as described in Sect. 3.2.

If we express the required confidence level of the estimation by Eq. (11), it can be seen that there are cases where not all observations in the entire time window are required for the output of the estimate. To take an extreme example, if the required confidence level is extremely small, such as $\delta = 0.01$, it is possible that Eq. (11) can be satisfied by the prior distribution alone without using any observations. Therefore, in this paper, the estimated aggregate value is output when it is determined that Eq. (11) is satisfied, even if it is in the middle of processing a time window.

In order to efficiently determine whether or not to output an estimate, we first describe the variance thresholds necessary for sufficiently accurate estimation. We then describe the computation of the estimation accuracy at the midpoint of the time window, and propose an output decision for the estimate based on stream processing with incremental Gaussian updates.

4.1   Threshold Value for Output Judgment of Estimated Value

In order to efficiently determine the validity of Eq. (11), the decision of the size of the confidence level is attributed to the decision of the size of the variance. Equation (11) determines whether each confidence level satisfies the required confidence level δ, but the calculation of the confidence level requires a range integral over a Gaussian distribution. Namely, it is inefficient to simply calculate the range integral for each device at each time.

Here, we note that the integration result does not depend on the expected value when using an integration range centered on the expected value in the Gaussian distribution [4]. In this case, Eq. (11) can be rewritten as follows:

$$\forall X \in \boldsymbol{X}, P(Y_X \in [-\epsilon, \epsilon]) \geq \delta \leftrightarrow \forall X \in \boldsymbol{X}, \int_{-\epsilon}^{\epsilon} \mathcal{N}(y \mid 0, \sigma_{Y_X}) dy \geq \delta. \tag{12}$$

This allows us to solve the following equation for the given required confidence level δ and error upper bound ϵ to find the *variance threshold* σ_θ required to meet the required estimation accuracy:

$$\int_{-\epsilon}^{\epsilon} \mathcal{N}(y \mid 0, \sigma_{Y_X}) dy = \delta. \tag{13}$$

Since it is a range integral centered on the expected value of the Gaussian distribution, the smaller the value of the variance, the larger the value of the confidence level. In other words, the following equation and Eq. (11) are equivalent:

$$\forall X \in \boldsymbol{X}, \sigma_{Y_X} \leq \sigma_\theta. \tag{14}$$

As described below, the variance of the aggregate value can be calculated incrementally by basic arithmetic operations, and thus the timing of the output can be determined more efficiently than a simple decision using range integration.

4.2 Incremental Update of Probabilistic Estimation

Using the variance thresholds described in the previous section, we can calculate the estimates and determine their output by determining the expected value and variance of the Gaussian distribution that the aggregate follows at each time in the time window. That is, for each sensor $X \in \boldsymbol{X}$, when the variance of the aggregate σ_{Y_X} is less than or equal to the threshold σ_θ, the expected value μ_{Y_X} is output as the estimated value. If we simply calculate the expected value and variance, duplicate calculations will be performed at each time. Thus, in the following, we describe an incremental update method suitable for stream processing.

First, we show the expected value and variance at any time $t^* \in [t, t + w)$ in a time window $\boldsymbol{x}^{[t,t+w)}$. Since $o^{[t,t^*]}$ is obtained as the observed value, the value of the device at each time can be estimated using the posterior probability distribution up to time t^* and the prior probability distribution at times after t^*. That is, from Eq. (8) and (9), the expected value of the aggregate $\mu_{Y_X}^{t^*}$ and variance $\sigma_{Y_X}^{t^*}$ at time t^* can be obtained by the following equation:

$$\mu_{Y_X}^{t^*} = \frac{1}{w} \left(\sum_{t' \in [t,t^*]} \mu_{X^{t'} | o^{t'}} + \sum_{t' \in (t^*, t+w)} \mu_X \right) \tag{15}$$

$$\sigma_{Y_X}^{t^*} = \frac{1}{w^2} \left(\sum_{t' \in [t,t^*]} \sigma_{X^{t'} | o^{t'}} + \sum_{t' \in (t^*, t+w)} \sigma_X \right). \tag{16}$$

Note that the prior distribution in the time window is constant.

We derive an asymptotic equation to incrementally update the expected value and variance. The following equations can be derived by transforming $\mu_{Y_X}^{t^*} - \mu_{Y_X}^{t^*-1}$ for the expectation value and $\sigma_{Y_X}^{t^*} - \sigma_{Y_X}^{t^*-1}$ for the variance:

$$\mu_{Y_X}^{t^*} = \mu_{Y_X}^{t^*-1} - \frac{\mu_X - \mu_{X^{t^*} | o^{t^*}}}{w} \tag{17}$$

$$\sigma_{Y_X}^{t^*} = \sigma_{Y_X}^{t^*-1} - \frac{\sigma_X - \sigma_{X^{t^*} | o^{t^*}}}{w^2}. \tag{18}$$

In other words, the Gaussian distribution of aggregate values can be updated by calculating the posterior probability distribution based on the new observed values o^{t^*} obtained at each time t^* and applying the difference.

5   Application to Fault Tolerance Assurance

In this section, we consider the estimation of aggregate values based on the probabilistic model described so far as a guarantee of fault tolerance in edge computing environments. Unlike existing distributed parallel stream processing systems that take fault tolerance into account, in an edge computing environment, there may be a situation in which communication with each device is disrupted due to a fault, making it impossible to retrieve and resend data observed during the fault. Namely, there may be cases where all the input data for a certain period of time is missing, such as time 5 and 6 in Fig. 3.

Even if all data for a certain period of time is missing due to a failure, the aggregate value can be estimated in the same way using our method. Although missing data due to a failure has a large impact, it is not theoretically different from partial missing data due to communication failure or specific missing data due to device failure. In other words, by estimating all data for all missing periods based on prior probabilities, we can calculate the estimated aggregate value and the confidence level reduced by the missing data.

Specifically, we use *checkpointing* together with the incremental probabilistic estimation described in Sect. 4. Checkpointing is a common method used in existing distributed parallel stream processing systems. The internal state of the process, i.e., the expected value and variance of the aggregate result, is periodically stored in a persistent storage area as a checkpoint. Then, when a failure occurs, the expected value $\mu_Y^{t_c}$ and variance $\sigma_Y^{t_c}$ of the latest checkpoint acquisition time t_c are restored to main memory and processing is restarted. However, as mentioned earlier, our method does not allow retransmission of lost data from the data source after recovery. All input data $o^{(t_c, t_d)}$ up to the recovery time t_d are assumed to have been lost and the process is restarted. If no failure occurs after that, the expected value μ_{Y_X} and variance σ_{Y_X} of the Gaussian distribution followed by the aggregate Y_X for sensor $X \in \mathbf{X}$ can be obtained as follows:

$$
\mu_{Y_X} = \frac{1}{w} \left(\sum_{t' \in [t, t_c]} \mu_{X^{t'} | o^{t'}} + \sum_{t' \in (t_c, t_d)} \mu_X + \sum_{t' \in [t_d, t+w)} \mu_{X^{t'} | o^{t'}} \right) \tag{19}
$$

$$
\sigma_{Y_X} = \frac{1}{w^2} \left(\sum_{t' \in [t, t_c]} \sigma_{X^{t'} | o^{t'}} + \sum_{t' \in (t_c, t_d)} \sigma_X + \sum_{t' \in [t_d, t+w)} \sigma_{X^{t'} | o^{t'}} \right). \tag{20}
$$

In the proposed method, there is no need to perform any special processing other than restoring the checkpoint as a failure recovery process. As shown in Eq. (15) and the second term in parentheses of (16), for any time $t^* \in [t, t + w)$ in a time window, the prior distribution is used to calculate the expectation and variance for times after t^*. Thus, at the time of the latest checkpoint acquisition t_c, we can say that the computation of the expected value and variance estimates for all missing data intervals (t_c, t_d) until the completion of the recovery has been completed. Therefore, after the failure is recovered, the approximate aggregation process with error guarantee can be resumed by simply restoring the latest checkpoint.

6   Related Work

We discuss the problem of missing data in time series data and related research on fault-tolerance guarantees for stream processing.

6.1   Missing Data in Time Series

The problem of missin data in time series has long attracted attention, and many studies have been reported. The basic approaches to the missing value problem are listwise and pairwise methods, which remove tuples and cases containing missing values [10]. Although these methods are widely used due to their simplicity, they have a problem of processing accuracy because the number of data to be processed is small and the number of data per attribute can be biased. Then, many methods have been proposed to estimate missing values with high accuracy, including interpolation by regression [6,11,13] and association rule mining [8].

For example, *stochastic regression imputation* takes into account the distribution of the data and randomly scatters the values from the regression line as the estimated values, and unlike conventional regression methods, it can provide more accurate estimates without underestimating the variance [6]. While this method targets the accurate estimation of each missing value, the proposed method aims to efficiently obtain an approximate aggregate result of the data stream containing the missing values. The proposed method takes into account the variance of the estimates through an error evaluation framework, and also allows for early output based on error guarantees and fault tolerance guarantees using checkpointing.

In addition, existing methods have various innovations to improve the processing accuracy, such as *multiple imputation* that combines multiple estimation results to solve the problems of underestimation of variance and estimation uncertainty [6]. However, there are some issues in low latency and fault tolerance, which are important in edge computing environments.

6.2   Fault Tolerance Assurance for Stream Processing Systems

For the problem of data loss due to system failures, many stream processing systems, including Flink [1] and Spark Streaming [2], guarantee robust fault tolerance without errors. These systems, which assume a parallel processing environment in the cloud, can guarantee the output of the result of processing the input data once without excess or deficiency, regardless of the occurrence of a failure, by backing up the internal state by checkpointing and restoring the input data based on retransmission and reprocessing. However, especially in a single-node environment, a failure at the network edge may cause a breakdown in communication with each device, making it impossible to acquire and resend data observed during the failure. The redundant configuration of the system is essential for retransmission of missing data, but the redundant configuration for each edge machine is expensive and impractical.

In order to reduce the cost of fault tolerance guarantee, Huang et al. [7] proposed an approximate fault tolerance method that guarantees the reliability of the processing results. This method reduces the amount of data to be backed up and the frequency of backups by backing up only when the threshold values for the number of unprocessed data and the amount of change in the internal state are exceeded, thereby improving the throughput. However, this method requires a detailed preliminary analysis that takes into account the actual node configuration in order to set a threshold value suitable for a Service Level Agreement (SLA), which poses many difficulties in actual operation.

In addition, all of these methods target only missing data due to stream processing system failures, and cannot handle missing input data due to sensor device failures or communication failures.

7 Conclusions

In this paper, we outlined our research on approximate aggregate processing methods that improve processing low latency, high reliability, and fault tolerance for edge computing environmental sensing applications. The proposed method uses a probabilistic model to estimate missing data due to sensor device failures or communication failures, and theoretically guarantees an upper bound on the error for aggregated results. Furthermore, we reduce the delay by outputting an approximate aggregate value that is incrementally computed at the shortest time when the error guarantee of our method satisfies the user requirement. In addition, our method can efficiently recover from failures even in situations where acquisition and retransmission of input data are impossible due to failures.

One of the future challenges is to follow the changes in correlation over time by updating the model dynamically. Dynamic model updating is expected to improve the accuracy of the process, since it will be able to take into account changes in correlations by time of day and season. However, in order to enable dynamic model updating, there are many issues that need to be considered in terms of consistency of processing results and fault tolerance, such as the adequacy of guaranteeing errors in aggregate results in response to model changes, and considering the effects of model update information loss due to node failures. For example, incremental processing and internal state management will be necessary to ensure that the theoretical error guarantee does not break down in response to changes in the model, including cases where the models are different before and after failure recovery. Adaptive checkpointing can also be considered, taking into account the trade-off between the degradation of processing performance due to the loss of model update information and the checkpointing cost, but it should also be balanced among multiple time windows with partial overlap, such as a sliding window.

Acknowledgement. This study is based on results obtained from a project, JPNP16007, commissioned by the New Energy and Industrial Technology Development Organization (NEDO). In addition, this work was supported partly by KAKENHI (16H01722, 20K19804, and 21H03555).

References

1. Apache Flink: Stateful Computations over Data Streams. https://flink.apache.org/
2. Spark Streaming | Apache Spark. http://spark.apache.org/streaming/
3. Bishop, C.M.: Pattern Recognition and Machine Learning. Springer, Boston (2006). https://doi.org/10.1007/978-1-4615-7566-5
4. Deshpande, A., Guestrin, C., Madden, S.R., Hellerstein, J.M., Hong, W.: Model-based approximate querying in sensor networks. VLDB J. **14**(4), 417–443 (2005)
5. Duffield, N., Xu, Y., Xia, L., Ahmed, N.K., Yu, M.: Stream aggregation through order sampling. In: Proceedings of the 2017 ACM on Conference on Information and Knowledge Management (CIKM). pp. 909–918 (2017)
6. Enders, C.K.: Applied Missing Data Analysis. Guilford Press, New York (2010)
7. Huang, Q., Lee, P.P.C.: Toward high-performance distributed stream processing via approximate fault tolerance. Proc. VLDB (PVLDB) **10**(3), 73–84 (2016)
8. Jiang, N., Gruenwald, L.: Estimating missing data in data streams. In: Proceedings of International Conference on Database Systems for Advanced Applications (DASFAA). pp. 981–987 (2007)
9. Johnson, T., Muthukrishnan, S., Rozenbaum, I.: Sampling algorithms in a stream operator. In: Proceedings of ACM SIGMOD. pp. 1–12 (2005)
10. Raymond, M.R., Roberts, D.M.: A comparison of methods for treating incomplete data in selection research. Educ. Psychol. Measur. **47**(1), 13–26 (1987)
11. Ren, X., Sug, H., Lee, H.: A new estimation model for wireless sensor networks based on the spatial-temporal correlation analysis. J. Inf. Commun. Converg. Eng. **13**(2), 105–112 (2015)
12. Takao, D., Sugiura, K., Ishikawa, Y.: Approximate streaming aggregation with low-latency and high-reliability for edge computing. IEICE Trans. Inf. Syst. **J104-D**(5), 463–475 (2021). (in Japanese)
13. Yi, X., Zheng, Y., Zhang, J., Li, T.: ST-MVL: Filling missing values in geo-sensory time series data. In: Proceedings of the 25th International Joint Conference on Artificial Intelligence (IJCAI), pp. 2704–2710 (2016)

Internet of Things Integrated with Multi-level Authentication for Secured IoT Data Stream Through TLS/SSL Layer

S. Yuvaraj[1] , M. Manigandan[2] , Vaithiyanathan Dhandapani[2](✉) , R. Saajid[1], and S. Nikhilesh[1]

[1] SRMSIT, Kattankulathur, Tamilnadu, India
yuvarajs@srmist.edu.in
[2] National Institute of Technology Delhi, Delhi, India
{manigandan,dvaithiyanathan}@nitdelhi.ac.in

Abstract. With the advancement in recent technology, the world is connected to multiple devices or peripherals through the Internet of things (IoT). Therefore, there is a need for proper configuration with the required authentication and secured data is required to transfer from a particular IoT device to the respective IoT cloud server configured to a specific application and IoT platform. Hence, this paper intends to configure any IoT device that integrates with a multi-level authentication for a secure IoT data stream over TLS/SSL. The multi-level authentication includes the first stage that includes cryptography-based fingerprint authentication to verify whether or not a particular IoT device is paired with the particular IoT cloud platform. In the second stage, each IoT device is configured with an API Key (Application Programming Interface) to enable data transfer from API enabled IoT device to the appropriate API Key activation of the IoT cloud platform. Multi-level authentication, activated with fingerprint authentication and an API key, thus provide a secure IoT data stream between every IoT device and the respective IoT cloud platform. To accomplish secure data transfer, we use a hashing algorithm such as Message Direct (MD5) cryptography for fingerprint authentication.

Keywords: IoT security and privacy · Cryptography · Symmetric cryptography · Hash functions · Message authentication

1 Introduction

The Internet of Things (IoT) is generally a network of physical components such as sensors connected to a microcontroller connected to built-in Wi-Fi modules (e.g. NodeMCU, Micropython etc.) is embedded or an external connecting device such as GSM module integrated with microcontrollers. Such wireless devices enable to exchange of real-time data for analysis with the configured IoT platform for data analysis. Nowadays, such IoT-enabled objects occupy various whitespace applications, including medical technology and manufacturing, to update the real-time data of things that are sent to the cloud server for specific embedded applications. As the number of physical objects increases,

© Springer Nature Switzerland AG 2022
S. Sachdeva et al. (Eds.): BDA 2021, LNCS 13167, pp. 245–258, 2022.
https://doi.org/10.1007/978-3-030-96600-3_18

data integrity and security play a crucial role in transferring secured data to the IoT cloud server. Figure 1 shows the physical things or objects like sensors to collect the real-time data and send the data through the internet gateways or satellites etc. Every physical object that is connected to the IoT platform via an IoT cloud server and all physical object data are aggregated in the IoT cloud server for data interpretation and analysis.

Fig. 1. Shows generalized IoT architecture

Figure 2 illustrates the generalized IoT device interaction with the client via the IoT platform and the server. Several authentications are performed, starting with the connection of the IoT device to the access point, from the access point to the ISP, from the ISP to the IoT server, and finally to the client device, to the real-time data from various IoT. As shown in Fig. 1 and Fig. 2, a system has the possibility of a vulnerability that can originate from IoT devices or IoT servers to the client devices or the Internet provider. The attacks can originate from unconfigured IoT devices trying to access the IoT server or IoT platform, leading to possible vulnerable attacks. These attacks can be traced back to a software bug that attacks the entire IoT cloud server through the unconfigured IoT devices. The devices are provided with known public-private keys to protect the IoT platform from unauthorized or users who access the IoT configured for specific IoT devices. Access the platform.

Data communication revolves around the Open System Interconnection (OSI) layer of a device. To understand how data communication works, you need to know how the OSI layer works. As is known in the prior art, the OSI architecture has seven layers, from the physical to the application. Each layer has its specific role in the OSI model, enabling the user to access the computer interface between data transmission across different layers. In terms of the OSI model, on the physical link layer, most communication devices such as Wi-Fi GPRS are used to interact with the IoT device and in turn with the IoT platform. To enable communication and increase the connectivity speed, IPv4 and IPv6 are used. The Lightweight Presentation Protocol (LPP) is used in addition to the fourth packet protocol (X.25). The presentation layer is responsible for forwarding and receiving the data from the IoT device and the server, to secure the data, various encryption techniques such as the Secure Sockets Layer (SSL) protocol are used. The application layer in the OSI model interacts with the IoT platform via multiple protocols such as MQTT, Rest API and HTTPS.

In this paper, we will focus on improving the security of the device to protect third party access from the server. To achieve this, we use multiple authentications in the presentation layer with encryption techniques such as TSL / SSL with fingerprint authentication from MD5 to ensure a secure exchange of certificates between the IoT device and the Domain Name Service (DNS) of the IoT platform. After fingerprint authentication verification, the physical data is uploaded to the server for further data interpretation and simulated for data visualization and analysis.

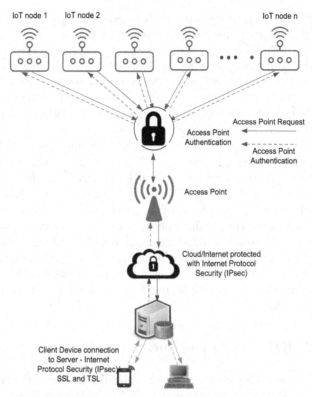

Fig. 2. Shows the general authenticated client device IoT device interaction with IoT platform via server

This paper is organized in the following order; Related work is discussed in Sect. 2 and Sect. 3 in Summaries of Security IoT Attacks and their Classifications. Section 4 describes the authentication and its type to secure the connection with the IoT device and the IoT platform. In Sect. 5, we discuss our methodology by looking at a real-time application. Section 6 discusses a case study on humidity and temperature monitoring and data transfer to the IoT cloud platform. In addition, the results are discussed in Sect. 7, followed by conclusions and future work.

2 Related Works

The Internet of Things, data integrity and security play an essential role in this increase in devices for application usage. Secure Sockets Layer (SSL) [1] and its successor Transport Layer Security (TLS) [2] are Internet protocols that enable secure networking. SSL/TLS [1] and [2] establish a secure connection between a client and a server over the Internet for data transmission. Concerning [1] and [2], application layer protocols provide a variety of capabilities, including data privacy, integrity, and user and system authentication services. The HyperText Transport Protocol Secure (HTTPS) [3] and the Secure Simple Mail Transfer Protocol (SSMTP) [4] are two application layer protocols that rely on SSL/TLS to communicate with the server and higher data communication between clients and server transfer. As a result, SS / TSL is seldom used to provide data authentication between a client and a server, as indicated in [1] and [2] the data transmission is required. Website Finger Printing (WFP) is a mechanism by which an observer uses statistical traffic analysis to identify the destination of the web transaction. [5]. WFP's uses include preventing enterprise system administrators from accessing prohibited websites and providing Quality of Service (QoS). When communication is encrypted and a client uses tunnelling and/or proxy services to mask its destination website addresses, WFP becomes very difficult. The application layer with SSL and TSL is required to offer handshaking between an IoT device and a client and server, as shown in [6]. According to the researchers in [7], an IoT SENTINEL can automatically detect the type of devices connected to an IoT network, reducing the vulnerability to attack. Further research into HTTP network traffic has been carried out by [8], which hinders legitimate network monitoring, traffic analysis and network forensics. For example, in cybercrime investigation the data utilizes a dictionary of SSL/TLS cipher suite and HTTP user agents [9]. In addition, the users are identified by their operating system, browser, and application they were using. According to [10], the system can classify encrypted network traffic and infer the user's operating system, browser, and application on his desktop or laptop computer.

3   Security - IOT Attacks and Their Classification

Every generalized IoT device in Fig. 1 can be further discussed in different phases such as Data aggregation, Storing the data in the cloud, Data Processing and visualization, Data transmission between an IoT device to the IoT server and to IoT Platform, and Client device interaction with the IoT platform to retrieve the data. Various types of sources of attacks associated with each phase and its possible damages on each stage are discussed in the following Table 1.

Table 1. Types of sources attacks on each phase and its major damages

Data perception	Data leakage, breach and authentication
Data storage	Denial of service, Attacks on data availability, access control and inter
Data processing	Authentication attack
Data transfer	Channel security, Routing protocols
Data delivery	Man to machine

The above Table 1 are again further categorized based on the attacks on OSI layer models, as shown in Table 2.

Table 2. OSI layer attacks

Application Layer	Revealing data, data destruction, user authentication
Transport Layer	DoS, heterogeneous and homogeneous network, Distributed DoS
Network Layer	Routing Protocol and Information
Physical Layer	External attack, link layer attack, access control

As mentioned in Table 1 and Table 2, appropriate security needs to be provided to ensure secure IoT deployment. Each IoT device might follow important terminologies such as authentication, authorization, integrity, confidentiality, data availability, and privacy. For a secured data transfer, various cryptographic techniques are used to enable data protection, process, and sharing of data. This paper intends to focus on cryptographic techniques such as MDH and SHA to generate a fingerprint for a particular IoT platform. The generated fingerprint is linked in the IoT device so that the fingerprint configured device can interact with that IoT platform. In the following Section IV, we discuss the methodology for secured data transfer with TSL/SSL certification enabled with fingerprint hash keys.

4 Authentication

This paper will discuss encrypted HTTPS - HTTP over SSL/TLS protocols for authentication and data transfer. The encrypted communication of the server is ported to the IoT device so that the server can first agree on the agreement of the encrypted key value. SSL/TLS fingerprint between IoT device and IoT cloud platform handshake varies between regular client-server interaction. Therefore, authentication occurs with the encrypted HASH key, which is generated using cryptographic techniques called fingerprint keys to ensure that every IoT channel is secure. In addition, device identification is done using an API key so that the server and device identify and validate each other to ensure that only the authorized devices/users access the protected resources.

4.1 Token-Based Authentication

A token is a string of characters generated by an authentication server that can identify a device and its access rights to a protected resource. Typically, the device sends a secret credential that is recognized by the server. When the certificates have been validated, the device is granted a token that can be used to access protected services. OAuth is one such industry-standard protocol. OAuth is a token-based authorization standard typically used in conjunction with an authentication method to ensure resource security.

4.2 Fingerprint Authentication

Certificates can be used in the authentication pipeline to identify devices uniquely, and the server, much like tokens can be used. The gadget can locate and verify the server when SSL is used in communication. The main goal of certificate-based authentication is to generate a certificate for each device and use it in the server to identify and authenticate it. This certification allows mutual authentication with SSL and certificate-based authentication to be set up. The server can recognise the device using the device's identity certificate. The device can identify the server using SSL certificates, and this creates a highly secure communication channel that is ideally suited for data transmission in companies. The SCEP protocol is one of the most widely used means of implementing certificate-based authentication; Let's look at what it is and why it matters. Based on the previous description, the device has an identity certificate to identify itself to the IoT server that the server trusts at the end of the process. When communicating with the server, the device identifies itself by sending a certificate as a header, and the message content is often signed with the device identity's private key. The server can check whether the device actually signed the content because it has the associated public key. The device can now securely access the protected APIs. However, security planning must be a cornerstone of the design and development process for IoT-related devices and services to ensure that they can connect and communicate securely and withstand accidental and deliberate attempts to undermine their integrity.

As with other sensitive online transactions, communication between IoT devices, applications, and back-end services should be encrypted using the Secure Socket Layer/Transport Layer Security Protocol (SSL/TLS). IoT management interfaces and applications must be designed to withstand brute force authentication attacks and set a high bar on user and administrator provisioning to prevent simple account compromises such as standard administrator credentials (or password-rate attacks).

Message Digest (MD5)
MD5 (Message Digest) is one of the most popular cryptographic hash functions. From Fig. 3, the predecessors of this function are the previous version of the algorithm. This function generates a 128-bit hash value or a 16-bit key called a fingerprint, and it has a 512-bit block to encrypt the code for security reasons. This code was mainly used as a tool to verify data integrity. The previous versions had serious security disadvantages because brute force attacks could crack them. The main purpose of the algorithm was to check the integrity of files being transferred over a network. The duplicates of the same files can be identified using a pre-calculated checksum. The MD5 accepts a random number of inputs to obtain a fixed output using permutation blocks. The next application found in MD5 was verification with digital signature authentication. Taking the previous version into account, MD4 was faster than MD5, but with some security drawbacks. The consideration of MD5 was therefore better.

Step 1 – Append Padding Bits. The primary step of the algorithm is padding, and padding is done to concatenate the input bits to a constant length so that the output remains constant. The primary purpose of padding is to expand the incoming bits. The message is elongated until the size is 64 bits smaller than the block size, so random bits are introduced to scramble the incoming message. It is 448 modulo 512 in size. Padding is done by adding a single 1 bit to the message and then appending 0 bits to 448. Now that the pad bits are added, the next step is to modify the length of the algorithm.

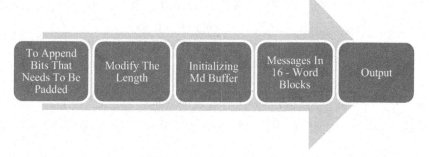

Fig. 3. Shows the generalized message direct MD5 block diagram

Step 2 – Append Length As mentioned earlier, the algorithm will now pad the length. A 64-bit representation of the message before adding the filler bits is repeated and added after step 1. If the resulting message is greater than $2^{\wedge}64$, then only lower-order 64 bits are used. Now the resulting bits are a multiple of 512 bits. It is also an exact length of 16 words that are encoded on the output.

Step 3 – Md Buffer. The 16 – bit word is taken and then split into four, four-word buffer. Each word is assigned to a 32 – bit word register named X, Y, Z, and W.

Step 4 – to Process the Message in 16 – Word Blocks. After the words, auxiliary functions to scramble the incoming message are taken. Each input takes three 32 – bits and produces one 32 – bit.

$$F (A, B, C) = (AB) \vee ((\sim(A)) \, C)$$
$$G (A, B, C) = AC \vee B \sim(C)$$
$$H (A, B, C) = A \oplus B \oplus C$$
$$I (A, B, C) = B \oplus (A \vee \sim(C))$$

Here the F function acts as a selector for the A, B, C bits. In the above function, the bits will not be in the same position one so the value will be the same. Note that ABC is independent and undistorted. Then the selected function goes to the remaining auxiliary values and acts bit by bit in parallel in order to generate an independent and undistorted value for the output again. In this case, the H function serves as a parity checker, which makes it possible to check any two errors. A 64-element table T [1 ... 64] is created with the sine function in this phase. The ith member of the table is denoted by T [i], which is equal to the integral component of 4294967296 times abs (sin (i)), where I is given in radians. The elements of the table are given in the appendix.

$$\text{For } i = 0 \text{ to } N/16\text{-}1 \text{ do}$$
$$\text{For } j = 0 \text{ to } 15 \text{ do}$$
$$\text{Set } X[j] \text{ to } M[i*16+j].$$
$$\text{End}$$
$$a=X; b=Y; c= Z \text{ and } d=W.$$

Let [abcd k s i] denote the operation:

ROUND 1

$$a = b + ((a + F(b, c, d)) + X[k] + T[i]) <<< s$$

[XYZW 0 7 1] [WXYZ 1 12   2] [ZWXY 2 17 3] [YZWX 3 22 4]
[XYZW 4 7 5] [WXYZ 5 12   6] [ZWXY 6 17 7] [YZWX 7 22 8]
[XYZW 8 7 9][WXYZ 9 12 10][ZWXY 10 17 11][YZWX 11 22 12]
[XYZW 12 7 13][WXYZ 13 12 14][ZWXY 14 17 15][YZWX 15 22 16]

ROUND 2

$$a = b + ((a + G(b, c, d) + X[k] + T[i]) <<< s)$$

ROUND 3

$$a = b + ((a + H(b, c, d) + X[k] + T[i]) <<< s)$$

ROUND 4

$$a = b + ((a + I(b, c, d) + X[k] + T[i]) <<< s)$$

[XYZW 0 6 49] [WXYZ 7 10 50] [ZWXY 14 15 51] [YZWX 5 21 52]
[XYZW 12 6 53] [WXYZ 3 10 54] [ZWXY 10 15 55] [YZWX 1 21 56]
[XYZW 8 6 57] [WXYZ 15 10 58] [ZWXY 6 15 59] [YZWX 13 21 60]
[XYZW 4 6 61] [WXYZ 11 10 62] [ZWXY 2 15 63] [YZWX 9 21 64]

The above is iteration carried out till I = 1, 2, 3, 4, and the letters are scrambled.

STEP 5 – OUTPUT

The output of the result is displayed as a 16-word output, which is generated as ABCD at the end. From the above steps, the output may not be the same even for the same input at different times. The MD5 algorithm has been thoroughly examined for errors.

SHA – Secure Hash Algorithm

The National Institutes of Standards and Technology (NIST), with the help of the government and private organizations, have compiled a collection of cryptographic algorithms. Given the need for stricter online security standards for businesses and the general public, this secure algorithm or encryption approach has helped address key cybersecurity issues. The SHA-0 algorithm, which uses 16-bit cryptographic keys, was one of the first secure hash algorithms to be developed. SHA-2 was developed in response to the need for better encryption standards in businesses. With 256 and 512 bit cryptographic keys, the SHA-2 uses a set of two functions. These algorithms are one-way functions, which means that it is almost impossible to change them back to their original data after they have been converted to their hash values. It transforms data using hash functions, which is a type of algorithm that includes bitwise operators, modular additions, and compression functions. The function is shown as a graphic representation as shown in Fig. 4.

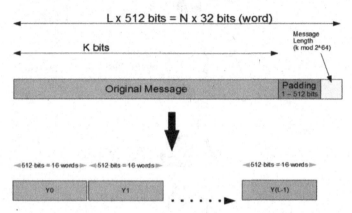

Fig. 4. Illustrates the pictorial representation of the SHA algorithm

It is computationally impossible to generate two messages with the same message digest. With a message digest, it is impossible to duplicate a message. Data Integrity Testing and Comparisons Finally, customers and business partners must understand the scope and impact of adverse events in the field by logging changes and activities on endpoints and management servers.

5   System Architecture

`In this paper, we will focus on improving the security of the device to protect third party access from the server. Implementation of multiple authentications in the presentation layer with encryption techniques such as TSL/SSL with fingerprint authentication from MD5 ensures a secure exchange of certificates between the IoT device and the IoT platform's Domain Name Service (DNS). After fingerprint authentication verification, the physical data is uploaded to the server for further data interpretation and simulated for data visualization and analysis.

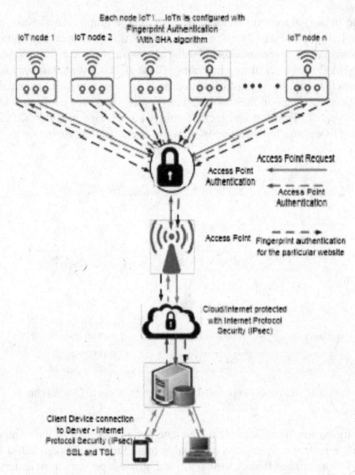

Fig. 5. Illustrated the generalized IoT architecture with Fingerprint authentication

To achieve the authentications at multiple levels, we intend to modify Fig. 2. The generalized IoT device interaction with the client via the IoT platform and the server with fingerprint authentication in each node is done using the SHA algorithm to generate a key for the respective IoT platform URL. The fingerprint authentication is sent along with the API key, so the real-time sensor data is only sent to the IoT platform for data visualization. The fingerprint sensor data can only be recorded in the IoT platform if verified by the fingerprint authentication between server and client.

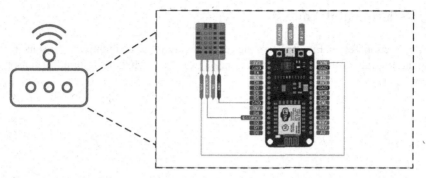

Fig. 6. Real time humidity and temperature monitoring using NodeMCU and Thingspeak configuration

Figure 5 avoids the possibility of a security breach originating from IoT. The attack could primarily originate from an unconfigured IoT device trying to access the IoT server and Access the IoT platform, leading to possible vulnerable attacks. These attacks can be traced back to a software bug that attacks the entire IoT cloud server through the unconfigured IoT devices. This article aims to look at real-time humidity and temperature monitoring using NodeMCU and Thingspeak configuration, as shown in Fig. 6. In addition, the whole process of multiple authentications for real-time sensor data visualization is shown in the Fig. 7.

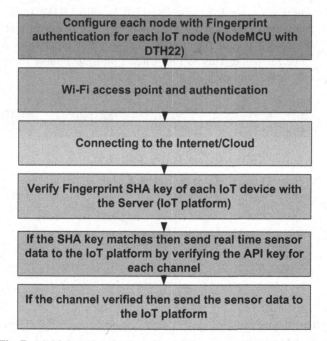

Fig. 7. Multiple authentications for real-time sensor data visualization.

6   Experimental Analysis

In this section below, Fig. 8 shows the real-time humidity and temperature sensor data captured in the Arduino CCS serial monitor from NodeMCU and humidity sensor, as shown in Fig. 6.

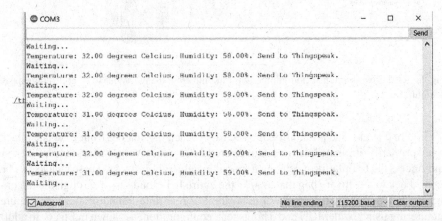

Fig. 8. Shows the real-time humidity and temperature sensor data acquired in serial monitor of Arduino CCS from NodeMCU

After checking each node IoT node (NodeMCU with DTH22) with fingerprint authentication configured with the IoT cloud platform, Fig. 9 shows the real-time sensor temperature and humidity measurement with Thingspeak. In order to receive the data, the API key must be compared. This API Key makes the system more secure. Figure 10 also shows that the device authentication does not match.

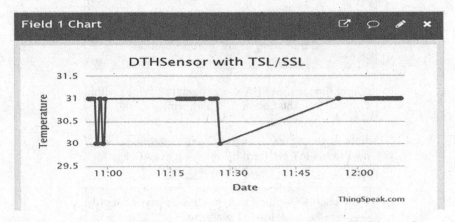

Fig. 9. Illustrates the real-time sensor temperature measurement with Thingspeak.

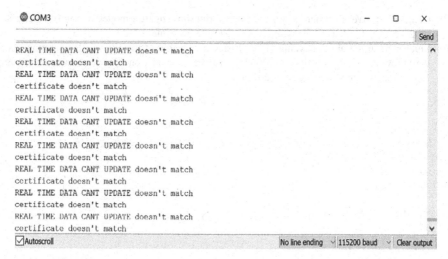

Fig. 10. TSL/SSL certificate authentication for IoT Real-time sensor module with Thingspeak.

7 Conclusion

From the analysis, the security for the device to protect third-party access from the server is achieved by configuring TSL / SSL with fingerprint authentication from MD5 to ensure a secure exchange of certificates between the IoT device and the IoT platform Domain name guarantee service (DNS). After fingerprint authentication verification, the physical thing data is uploaded to the server for further data interpretation and simulated for data visualization and analysis.

References

1. The Secure Sockets Layer (SSL) Protocol Version 3.0. No. RFC 6101 (2011)
2. The Transport Layer Security (TLS) Protocol Version 1.2, No. RFC 5246 (2008)
3. HTTP over TLS, no. RFC 2818 (2000)
4. SMTP service extension for secure SMTP over Transport Layer Security, no. RFC 3207 (2002)
5. Herrmann, D., Wendolsky, R., Federrath, H.: Website fingerprinting: attacking popular privacy enhancing technologies with the multinomial Naive-bayes classifier. In: Proceedings of the 2009 ACM Workshop on Cloud Computing Security, New York (2009)
6. Anjali, T., Krishna, S., Poonam, Z.: Comparison between SSL and SSH in network and transport layer. Int. J. Comput. Sci. Inf. Technol. **6**(6), 5432–5434 (2015)
7. Miettinen, M., Marchal, S., Hafeez, I., Asokan, N., Sadeghi, A.-R., Tarkoma, S.: IOT SEN-TINEL: Automated device-type identification for security enforcement in IoT. In: 2017 IEEE 37th International Conference on Distributed Computing Systems (ICDCS) (2017)
8. Husák, M., Čermák, M., Jirsík, T., Čeleda, P.: HTTPS traffic analysis and client identification using passive SSL/TLS fingerprinting. EURASIP J. Inf. Secur. **2016**(1), 1–14 (2016). https://doi.org/10.1186/s13635-016-0030-7

9. Ghaleb, T.A.: Website fingerprinting as a cybercrime investigation model: role and challenges. In: 2015 First International Conference on Anti-Cybercrime (ICACC) (2015)
10. Muehlstein, J., et al.: Analyzing HTTPS encrypted traffic to identify user's operating system, browser and application. In: 2017 14th IEEE Annual Consumer Communications & Networking Conference (CCNC) (2017)

Enhanced Bond Energy Algorithm for Vertical Fragmentation of IoT Data in a Fog Environment

Parmeet Kaur[✉]

Jaypee Institute of Information Technology, Noida, India
Parmeet.kaur@jiit.ac.in

Abstract. The vast benefits of Internet of Things (IoT) stem from the big data collected by the IoT devices. Currently, majority of this data is transferred to cloud-based servers where it is processed and analyzed for further actions. However, the transfer of the IoT big data to the cloud utilizes high network bandwidth and may also result in unacceptable response time for time-critical applications. The advent of fog computing has provided a direction for addressing these constraints of bandwidth and latency. IoT data related to time-critical applications can be stored at the fog nodes for faster processing. However, this data is generally high dimensional apart from being large sized and thus needs to be efficiently stored for scalability of related applications. Fragmentation of data has been established as a suitable technique for scaling of big data storage. This work presents an enhanced version of the existing Bond Energy Algorithm for vertical fragmentation of IoT data to be stored and processed on fog nodes. The proposed algorithm considers the criticality, access frequency and latency requirements of applications to determine how data may be fragmented across the fog nodes. The results of the algorithm show a marked improvement in performance as compared to the scenario where data is not fragmented. The findings are applicable to both relational and non-relational databases.

Keywords: Internet of Things · Big data · Fog · Vertical fragmentation · Bond energy algorithm

1 Introduction

The Internet of Things (IoT) refers to the interconnection between physical objects that are equipped with sensors and related software for exchanging data with other devices or systems over the internet [1]. These devices are widely present in appliances of everyday use such as cars, refrigerators, washing machines, monitoring devices etc. as well as in industrial tools and machinery. The number of connected IoT devices has surpassed 10 billion today and expected to grow further.

The IoT devices continually sense and record data from their environment and share this data for further processing. The advancements in the technologies of cloud and mobile computing, big data, machine learning etc. have made it possible to extract valuable insights from the IoT big data. IoT data is transferred to cloud-based servers where

© Springer Nature Switzerland AG 2022
S. Sachdeva et al. (Eds.): BDA 2021, LNCS 13167, pp. 259–267, 2022.
https://doi.org/10.1007/978-3-030-96600-3_19

it is later stored and analyzed [2]. However, IoT data is generally high dimensional and noisy which makes it challenging to store it efficiently. Many time-critical applications may also require this data be analyzed in real-time. Further, some applications may have privacy or security concerns in transferring data to the cloud. Fog based architectures are a recent advancement that can support all such applications.

Fog computing adds a layer in the IoT-cloud continuum. It provides storage and computing power close to the source of data, i.e., the IoT nodes [3]. On one side, the resource limitations of IoT devices can be overcome using the fog nodes. On other side, the fog nodes being proximate to the IoT nodes can help in reducing the latency caused by transfer of data to and from the cloud. This makes fog computing a viable alternative to cloud for applications that require low latency. It is also appropriate for applications whose data cannot be transferred to cloud for privacy or security reasons. Cloud computing is used in fog architecture for long term data storage or for analytics with non-critical response time requirements.

As compared to the cloud servers, fog nodes do not have 'infinite' compute or storage capabilities. Therefore, it is imperative that resources on these nodes be used efficiently. This work focusses on a technique for efficient and scalable IoT data storage at fog nodes. Specifically, we present an enhanced version of the existing Bond Energy Algorithm (BEA) [4] that can be used to vertically fragment the multi-dimensional IoT data onto the fog nodes. Vertical fragmentation is a well-established method for improving data access time by grouping data attributes frequently accessed together in dedicated fragments. This reduces the number of accesses to irrelevant data; thereby, increasing data availability and decreasing the response time. The BEA was proposed for vertical fragmentation in relational databases. The enhanced version, presented here, is based on specific requirements of IoT data in a fog-cloud environment and can be applied in relational as well as non-relational databases. It considers the response time, access frequencies and criticality requirements of applications accessing the data for its fragmentation.

The rest of the paper is structured as follows: Sect. 2 outlines the related work in the problem under consideration. Section 3 describes the proposed vertical fragmentation algorithm. We discuss the results of implementation of the algorithm in Sect. 4 and subsequently, conclude the presentation.

2   Related Work

Value extraction from big data is dependent on a large extent on whether it is stored efficiently. It is important that data to be used for analytics is easily and readily accessible to the applications. The use of fog computing (or any form of edge computing) enriches the process of IoT data storage by providing a platform where any type of preprocessing of data can take place. This may include noise removal, privacy preservation, encryption or fragmentation of data before it gets stored. Any type of subsequent operation such as querying, machine learning, anomaly detection etc. can then take place with a low latency or even in real time.

Researchers have come up with various solutions for efficient data storage for big data such as IoT data. The work in [5] utilizes Hadoop Distributed File System (HDFS) for optimization of IoT data access storage algorithms in the cloud. The authors focus on the requirements of IoT data storage for configuring storage locations of data access information. They also optimize the IoT deployment topology and file storage. The presented technique is shown to improve the data read and write efficiency.

Authors of [6] have emphasized on the security aspects of cloud storage for IoT data. They provide ways of securing data during its collection, storage as well as accessing it. The work highlights the limitations of cloud storage. The challenges of dealing with large volumes of structured and unstructured data have been discussed in [7]. The work presents a cloud-based software defined storage system that is suitable for multi-tenant software-as- a-service applications. The presented system considers multi-attribute data that needs to be stored with an aim to improve storage utilization and implement tenant isolation.

In order to address constraints of cloud assisted storage, the work in [8] presents a method to use edge and fog devices for distributed data storage. The authors have developed ElfStore, a federated data store that can store data on fog and edge devices and manage failures. The data store utilizes Bloom filters to locate data efficiently within the network and relies on replication for fault tolerance. Experiments have verified the reliability and scalability of the data store.

Another work that has leveraged fog computing for data storage and access has been presented in [9] for Industrial IoT (IIoT). A flexible and economic framework has been designed that considers latency requirements of applications to determine whether data should be stored on cloud or fog servers. to solve the problems above by integrating the fog computing and cloud computing.

The work in [10] emphasizes on the need to design an efficient distributed database for a good performance of the related applications. Fragmentation is highlighted as an important research problem along with allocation of these fragments on appropriate nodes. The BEA has been modified with a different affinity measure to cluster related attributes together. Similarly, the affinity measure has also been modified to improve the clustering of attributes in [11]. A scheme, called Vertical Fragmentation, Allocation and Replication scheme (VFAR) has been presented in [12]. Data is fragmented initially and then it is allocated as well as replicated at multiple locations for high availability. Another optimization technique for horizontal fragmentation and allocation has been put forth in [13]. The scheme aims to reduce the transmission costs of data allocation. A metaheuristic- based version of BEA has been presented in [14] that combines differential evolution algorithm with the BEA. The objective is to explore various combinations of attribute clustering for efficient fragmentation.

This work presents a vertical data fragmentation technique for IoT data to be stored on fog nodes. The technique is an enhanced version of Bond Energy Algorithm (BEA) that considers the requirements of time-critical applications. The algorithm presented in this work is different from the existing works as it is focused on the characteristics of IoT data and the applications storing or accessing this data.

3   Proposed Algorithm

IoT data is usually multi-dimensional and storing it at one node is not a viable option. This is because any given application will need to access only a given subset of attributes from this data. For example, a smart home may be generating data from smoke and fire sensors, temperature and humidity sensors, smart meters, smart thermostat, video doorbell, door and window opening or closing sensors, motion sensors etc. Of these data, an anomalous activity detection algorithm may require data from video doorbell and window or door sensors. A fire detector may fetch data from smoke and fire sensors only. Similarly, other applications may also use only specific data. If the entire data is stored at one location within the same data schema, it will lead to scalability issues.

This work proposes a vertical fragmentation approach to determine which subsets of the total data attributes should be grouped together and stored at the same location. The classical Bond Energy Algorithm for vertical fragmentation of relational databases is modified to make it suitable for IoT data storage in the fog environment. The classical BEA is described next before discussing the enhanced BEA.

3.1 Classical Bond Energy Algorithm

The bond energy algorithm (BEA) was developed to determine how attributes within a database may be grouped for efficient access by queries. The idea was to vertically fragment the database based on which attributes are used together by queries. The attributes with common usage are said to possess higher affinity or bond amongst them. If such attributes are placed together, queries will need to access one (or few) fragments instead of the complete database.

The steps involved in BEA are as follows:

1. Derive the attribute affinity (AA) matrix, whose rows and columns correspond to the given attributes, to represent the closeness or affinity between attributes. For the attributes that are accessed together, use Eqs. 1 and 2 to measure affinity.
2. Cluster attributes in the AA matrix such that those with high bonds are brought closer. The resultant is the clustered affinity (CA) matrix.
3. Fragment the attributes in CA such that each fragment F_i, where $1 \leq i \leq n$ and n is the number of fragments, contains a set of attributes accessed by a specific application (s).

The calculation of AA matrix is performed by finding usage of attributes by queries along with the measure of query accesses. Affinity of any two attributes a_m and a_n is calculated as:

$$Aff(a_m, a_n) = \sum_{all \ queries \ that \ access \ a_m \ and \ a_n} Query \ Access \qquad (1)$$

$$Query \ Access = \sum_{all \ sites} access \ frequency \ of \ a \ query \qquad (2)$$

The CA matrix is formed by placing the columns and rows of attributes in a way that maximizes the overall or global affinity (GA). GA is defined as:

$$GA = \sum_m \sum_n affinity \ of \ a_m \ and \ a_n \ with \ their \ neighbors \ in \ AAmatrix \qquad (3)$$

Placement of attributes in CA is evaluated by measuring the contribution of each placement as

$$cont(a_i, a_k, a_j) = 2 \ bond(a_i, a_k) + 2 \ bond(a_k, a_l) - 2 \ bond(a_i, a_j) \qquad (4)$$

where

$$bond(a_x, a_y) = \sum_{z=1}^{n} Aff(a_z, a_x) Aff(a_z, a_y) \qquad (5)$$

Once the attributes are clustered, the CA matrix is fragmented into a set of fragments where each fragment is made up of a set of continuous attributes $\{a_x, a_{x+1}, \dots a_{x+k}\}$ from CA such that no (or minimal) number of applications need to access more than 1 fragment.

The classical BEA focusses only on the concurrent access of attributes by applications. However, there are more parameters that can determine the affinity or togetherness of attributes in fragments. Therefore, the enhanced version of BEA endeavors to find an improved measure of affinity.

3.2 Enhanced BEA

The work presents an enhanced version of BEA considering the needs of IoT data that is to be stored on fog nodes. The data is multi-dimensional and is stored on fog nodes so that it can be accessed in real-time by time-critical applications. Further, various applications may each access only a subset of the attributes. With these considerations, we modify the measure of affinity between attributes as described next.

The attribute affinity measure between two attributes a_i and a_j of a dataset is with respect to the set of applications $P = (p_1, p_2, \dots, p_q)$. It is conventionally given as the number of times the two attributes are accessed together by any of the applications in P. These applications may be executing on more than 1 node and hence, accesses at all nodes are counted.

The Enhanced BEA modifies the measure of attribute affinity by giving importance to (a) time-criticality of the applications accessing the attributes, t_{pi} (b) Priority of applications accessing the attributes, pr_{pi}. This is in addition to the number of concurrent accesses by applications to these attributes.

$$Mod\_Aff(a_m, a_n) = \sum_{all \ queries \ that \ access \ a_m, a_n} all \ queries \ that \ access \ a_m, a_n \qquad (6)$$

$$Mod\_Query \ Access = \sum_{allnodes} [access \ frequency \ of \ a \ query]$$
$$*Sum \ of \ time - criticality \ of \ the \ applications \ accessing \qquad (7)$$
$$a_m \ and \ a_n * Sum \ of \ Priority \ of \ applications \ accessing \ a_m \ and \ a_n$$

Here, both the parameters t_{pi} and pr_{pi} for an application p_i are user-defined and their values lie between 0 and 1. A higher value implies the higher time-criticality or higher priority for the user. It is straightforward to see the implication of modifying the affinity measure. Consider an application p_1 that accesses the attributes a_1 and a_2 together for

50 times and an application p_2 that accesses the attributes a_2 and a_3 together for 50 times. However due to the high volume of data, it is required to fragment the data. Let p_1 be a time-critical and high priority application with values of both parameters being 0.8 each. Let p_2 not be a time-critical application and its priority is medium. The values of time-criticality be 0.3 and priority be 0.5. Therefore,

$$Mod\_Aff(a_1, a_2) = 50 * 0.8 * 0.8 = 32$$

$$Mod\_Aff(a_2,) = 50 * 0.3 * 0.5 = 7.5$$

The value of affinity measure is same for both applications (i.e., 50) when calculated using the classical BEA. This is because the classical BEA does not distinguish between the applications accessing the data. However, this is not feasible for modern applications using IoT data. For instance, p_1 may be a fall-detection application in a smart home while p_2 may be a batch analytics application for prediction of power usage. In such a case, the modified affinity measure will give a better performance for p_1 by reducing its access time. The complete Enhanced BEA is presented in Fig. 1.

Input: The attribute usage information by applications
Output: The clustered affinity matrix CA
Procedure:
1. Derive the Modified attribute affinity (AA) matrix, whose rows and columns correspond to the given attributes, to represent the closeness or affinity between attributes. Use equations 6 and 7 to measure modified affinity.
2. Cluster attributes in the AA matrix such that those with high bonds are brought closer. The resultant is the clustered affinity (CA) matrix.
3. Partition the attributes in CA such that each partition Pi , where $1 \le i \le n$ and n is the number of partitions, contains a set of attributes accessed by a specific application (s).

Fig. 1. Enhanced BEA

4 Implementation and Results

The enhanced BEA has been evaluated using 3 datasets, 2 from the IoT domain [15] and 1 random dataset, as listed in Table 1. Queries based on these datasets as well as the values of their criticality and priority have been randomly generated. It is assumed that each dataset is accessed by applications of 3 types, as described in Table 2. The evaluation of enhanced BEA has been performed by measuring the number of fragments for each dataset that need to be accessed by applications of each type (as in Table 2). The difference from the classical BEA is also discussed.

Figure 2a and b illustrate the Attribute Affinity matrices obtained for the dataset (with 9 attributes) using classical BEA and enhanced BEA. These matrices are further used to obtain 2 fragments each of the dataset. The fragments obtained by the classical BEA are (a_9, a_3, a_1, a_7, a_5) and {a_2, a_4, a_8, a_6} while those obtained by the enhanced

Table 1. Data sets

S No	Number of attributes used
1	4
2	6
3	9

Table 2. Type of applications.

Type	Priority and latency	Applications of this type
1	High	p_1, p_2
2	Medium	p_3, p_4
3	Low	p_5, p_6

BEA are $\{a_9, a_1, a_7, a_8, a_2\}$ and $\{a_3, a_4, a_5, a_6\}$. By observing the Attribute Affinity Matrix of Fig. 2a, we find that attributes with high affinity, based on access frequencies, are placed in same fragment since classical BEA only considers access frequency for fragmentation. For example, a_1, a_3 and a_5 have high attribute affinities and hence, they are placed in the same fragment. However, attributes a_1 and a_8 are placed in separate fragments since these have low attribute affinity measure.

However, the Enhanced BEA also considers priority and latency requirements of the applications accessing the data. The application p_2 accessing the attributes a1 and a8 in our experiments has a high priority and is time-sensitive. Therefore, these attributes are placed in the same fragment by modified BEA. Similarly, the applications p_5 and p_6 accessing attributes a_1 and a_3 have a low priority and are not latency-sensitive. Therefore, these attributes are placed in different fragments by the enhanced BEA Similar results have been obtained for other 2 datasets also.

Table 3 illustrates the number of fragments that need to be accessed by applications of each category for both classical and enhanced BEA. It can be observed that low priority and latency-insensitive applications need to access both fragments if enhanced BEA is applied, which is as expected. If at any time the number of accesses needs to be lowered, it can be done by increasing the priority and latency sensitivity of the application.

The proposed version of the BEA can be readily used in fragmentation of both structured and unstructured data. The algorithm only needs the attribute usage information from the applications for fragmenting the dataset. Hence, there is a wide scope for the application of the proposed enhanced BEA in IoT as well as other application domains.

	a_1	a_2	a_3	a_4	a_5	a_6	a_7	a_8	a_9
a_1	10	2	50	5	40	2	50	5	40
a_2	2	10	5	45	2	30	2	40	5
a_3	50	5	10	2	40	2	45	5	50
a_4	5	45	2	10	2	40	5	40	2
a_5	40	2	40	2	10	2	2	45	1
a_6	2	30	2	40	2	10	2	42	1
a_7	50	2	45	5	2	2	10	1	35
a_8	5	40	5	40	45	42	1	10	2
a_9	40	5	50	2	1	1	35	2	10

(a)

	a_1	a_2	a_3	a_4	a_5	a_6	a_7	a_8	a_9
a_1	10	1.8	0.5	4.05	0.4	1.8	0.5	4.5	0.4
a_2	1.8	10	4.5	45	1.62	40	1.62	40	4.5
a_3	0.5	4.5	10	2	0.4	2	0.45	5	0.5
a_4	4.05	45	2	10	2	40	4.5	40	1.62
a_5	0.4	1.62	0.4	2	10	2	2	45	1
a_6	2	40	2	40	2	10	2	42	0.9
a_7	0.5	1.62	0.45	4.5	2	2	10	1	35
a_8	4.5	40	5	40	45	42	1	10	2
a_9	0.4	4.5	0.5	1.62	1	0.9	35	2	10

(b)

Fig. 2. Attribute affinity matrix using (a) classical BEA (b) Enhanced BEA

Table 3. Number of fragments accessed by applications.

Type of application	Number of fragments that need to be accessed	
	Classical BEA	Enhanced BEA
High priority and latency sensitive	2	1
Medium priority and latency insensitive	2	1
Low priority and latency insensitive	1	2

5 Conclusion

The paper has presented an enhanced version of the classical Bond Energy Algorithm employed for vertical fragmentation of large IoT data sets in a fog environment. The modifications in BEA have considered the priority and latency requirements of applications accessing the data in an IoT system while fragmenting the data. This is in contrast to considering only the frequency of accesses by applications in classical BEA. The implementation of the enhanced BEA has shown that it is successful in achieving fragments such that high priority and time-sensitive applications do not need to access multiple data fragments. This is a significant requirement for multiple critical domains such as health care or industrial monitoring that need immediate data processing. An implementation of the enhanced BEA can result in efficient data storage at fog nodes.

References

1. Ashton, K.: That "internet of things" thing. RFID J. **22**(7), 97–114 (2009)
2. Ahmed, E., et al.: The role of big data analytics in Internet of Things. Comput. Netw. **129**, 459–471 (2017)
3. Mahmud, R., Kotagiri, R., Buyya, R.: Fog computing: a taxonomy, survey and future directions. In: Di Martino, B., Li, K.-C., Yang, L.T., Esposito, A. (eds.) Internet of Everything. IT, pp. 103–130. Springer, Singapore (2018). https://doi.org/10.1007/978-981-10-5861-5_5
4. Hoffer, A., Severance, D.G.: The use of cluster analysis in physical database design. In: Proceedings 1st International Conference on Very Large Databases, Framingham (1975)
5. Wang, M., Zhang, Q.: Optimized data storage algorithm of IoT based on cloud computing in distributed system. Comput. Commun. **157**, 124–131 (2020)
6. Wang, W., Xu, P., Yang, L.T.: Secure data collection, storage and access in cloud-assisted IoT. IEEE Cloud Comput. **5**(4), 77–88 (2018)
7. Sharma, A., Kaur, P.: TOSDS: tenant-centric object-based software defined storage for multi-tenant saas applications. Arab. J. Sci. Eng. **46**(9), 9221–9235 (2021). https://doi.org/10.1007/s13369-021-05743-z
8. Monga, S.K., Ramachandra, S.K., Simmhan, Y.: ElfStore: a resilient data storage service for federated edge and fog resources. In: 2019 IEEE International Conference on Web Services (ICWS), pp. 336–345. IEEE (2019)
9. Fu, J.-S., Liu, Y., Chao, H.-C., Bhargava, B.K., Zhang, Z.-J.: Secure data storage and searching for industrial IoT by integrating fog computing and cloud computing. IEEE Trans. Industr. Inf. **14**(10), 4519–4528 (2018)
10. Rahimi, H., Parand, F.-A., Riahi, D.: Hierarchical simultaneous vertical fragmentation and allocation using modified bond energy algorithm in distributed databases. Appl. Comput. Inform. **14**(2) 127–133 (2018)
11. Rahimi, H., Parand, F.-A., Riahi, D., 2015. Hierarchical simultaneous verticalfragmentation and allocation using modified bond energy algorithm in distributed databases. Appl. Comput. Inform. **14**, 127–133 (2015)
12. Abdel-Raouf, O.: A new hybrid flower pollination algorithm for solvingconstrained global optimization problems. Int. J. Appl. Oper. Res. **4**, 1–13 (2014)
13. Amer, A.A., Sewisy, A.A., Elgendy, T.M.A.: An optimized approach for simultaneous horizontal data fragmentation and allocation in distributed database systems (DDBSs). Heliyon 3, e00487 (2017)
14. Mehta, S., Agarwal, P., Shrivastava, P., Barlawala, J.: Differential bond energy algorithm for optimal vertical fragmentation of distributed databases. J. King Saud Univ. Comput. Inform. Sci. **34**, 1466–1471 (2018)
15. https://www.kaggle.com/atulanandjha/temperature-readings-iot-devices
16. https://www.kaggle.com/open-aq/openaq

Approximate Query Processing
with Error Guarantees

Tianjia Ni$^{(\boxtimes)}$, Kento Sugiura, Yoshiharu Ishikawa, and Kejing Lu

Graduate School of Informatics, Nagoya University,
Chikusa-ku, Nagoya-shi, Aichi 464-8601, Japan
{ni,lu}@db.is.i.nagoya-u.ac.jp, {sugiura,ishikawa}@i.nagoya-u.ac.jp

Abstract. In recent years, with the increase of data and the sophistication of analysis requirements, query processing in databases has become more important. Recently, approximate query processing (AQP) was proposed for efficiently executing database queries on big data. In this research, we focus on synopsis construction on a relational database and the query technology based on it, which is called Bounded Approximate Query (BAQ) proposed in 2019. BAQ is a synopsis construction method that focuses on aggregate queries using SQL, and realizes error-guaranteed query processing by grouping the dataset into the synopsis. In this paper, we point out the limitations of queries and datasets in BAQ and based on the result of experiments, we prove that the proposed method can be applied efficiently to data wider than the original BAQ with smaller synopsis within the error guarantee.

Keywords: Approximate query processing · Query processing · Aggregate estimation

1 Introduction

In recent years, query processing in databases has become more and more important due to the increasing amount of data and the sophistication of analysis requirements. Recently, **approximate query processing** (AQP) has attracted much attention as a technique for efficiently executing database queries on large amounts of data [1,2,6]. In this paper, we focus on the approximate query processing (AQP), which is a technique to efficiently perform query processing using not the whole database but a part of the database or summary data, although it is not accurate. However, since query results contain errors, it is important to reduce the error and to estimate the error. So for numerical data, approximate calculations are important.

Approaches to approximating numerical data can be broadly divided into two categories: those that do not retain any data and those that retain a portion of the data. Approximation methods using functions such as histograms and wavelets, and research [5,7,9,10] using statistical models can approximate large-scale data. However, since such methods do not retain the original data, the

© Springer Nature Switzerland AG 2022
S. Sachdeva et al. (Eds.): BDA 2021, LNCS 13167, pp. 268–278, 2022.
https://doi.org/10.1007/978-3-030-96600-3_20

obtained results are not easy to understand for users. On the other hand, there are some studies [3,4,8,9] on methods that retain part of the original data, such as sampling, quantile and BAQ.

Li et al. [3] proposed a method of constructing a synopsis on a relational database (BAQ) that can retain part of the data and guarantee the error of the query result. In query processing, a synopsis can be used to return error-guaranteed results for a given query. Since some of the original data is saved, it has the advantage that the user can easily grasp the data. In addition, compared to other approaches that use some of the data (sampling, quantile, etc.), BAQ can efficiently guarantee errors not only for the COUNT, SUM, and AVG functions, but also for the MIN and MAX functions.

However, there are some problems with the BAQ.

1. The BAQ can guarantee errors only for positive numeric data. Besides, When considering positive numeric data in BAQ, it is concluded that the error in calculating the MIN/MAX function is small and can be ignored, but considering negative numeric data, there is an estimation error that cannot be ignored.
2. Although the size of the synopsis generated by BAQ is smaller than the original data samples, there is still space for improvement in the algorithm for generating the synopsis in order to reduce it further.
3. The BAQ can only support the simply query. For real business case, approximate query processing techniques that can support complex query for the relational database are needed.

In order to overcome the above limitations, we propose an error-guaranteed approximate query processing framework, which is an extension [3]. Given a query set that represents a user-defined error threshold and a typical workload, we generate a synopsis from the database by offline processing and use the synopsis to answer an efficient online query.

In this paper, we propose the better buckets partitioning method in Sect. 3.1 to generate the smaller synopsis than BAQ. In Sect. 3.2 and Sect. 3.3, we extend our method to support the complex query which is defined in Sect. 2.1. By our proof [11], the approximate query processing framework can provide error guarantees on real data. The results in Sect. 4.2 can also show that the proposed method is more accurate and fast than BAQ.

2   Preliminary

In this chapter, we explain the concepts necessary for the discussion in this paper. In the following, we discuss the data and queries to be covered, the relative error as a measure of error, and bucket partitioning method of numerical attributes.

2.1   Queries

In this study, we consider a single relation schema $R = (A_1, A_2, ..., A_m)$ and treat a table T with n tuples as its instances. We assume that each $A_i \in R$ is a numerical or categorical attribute, and denote its domain by D_i. For the above table T we assume an OLAP query that uses the following operations.

- Joins between tables (including self-joins)
- Aggregation by COUNT, MIN, MAX, SUM, and AVG.
- Selection using $=$, $\neq$, EXIST for categorical attributes, and $=$, $\neq$, $>$, $\geq$, $<$, and $\leq$ for numeric attributes.
- Selection using logical OR $\vee$, logical AND $\wedge$, and negation $\neg$ for the above conditions.
- Grouping for categorical attributes.

For example, the following **Query** 1 is treated as a processing target.

```
SELECT c_custkey,c_name,
       SUM(l_price*(1-l_discount)) AS revenue,
       c_acctbal, n_name, c_address,c_phone
FROM customer, orders, lineitem, nation
WHERE o_orderdate >= '1993-10-01'
  AND o_orderdate < '1994-01-01'  AND l_returnflag = 'R'
  AND c_custkey = o_custkey  AND l_orderkey = o_orderkey
  AND c_nationkey=n_nationkey
GROUP BY c_custkey, c_name, c_acctbal, c_phone, n_name, c_address
ORDER BY revenue DESC
```

2.2   Relative Error

In this study, we consider the relative error err as the error between the true value and the approximate value of the aggregate result for a given OLAP query. The relative error of two values $x, y \in \mathbb{R}$ is determined by the following equation.

$$err(x,y) = \begin{cases} \frac{|x-y|}{\epsilon} & (x = 0) \\ \left|\frac{x-y}{x}\right| & (\text{otherwise}) \end{cases} \tag{1}$$

Where x expresses the true value, y expresses the approximate value, and $\epsilon \in \mathbb{R}$ is a very small positive value.

In this study, we assume that the user provides a set of queries Q, which represents the processing workload and the relative error threshold δ. Working under this assumption, for the aggregate calculations in the query set Q, we generate a set of reduced tuples by partitioning the numerical values into the buckets that keep the relative error of approximation within δ. This reduced set of tuples is called a **synopsis** in this research.

3   Synopsis Generation and Query Processing

In this chapter, we explain how to generate synopsis based on Sect. 3.1 and process queries in order to efficiently handle queries with join processing and complex decision conditions.

3.1 Bucket Partition

We describe the bucket partitioning of numerical attributes and the construction of the synopsis, which is important for the error guarantee. The bucket set B_i is a set of non-overlapping ranges of a numerical attribute $A_i \in R$, generated so that the relative error of any two values in each bucket is within δ. In other words, for each bucket $b \in B_i$, the following equation holds.

$$\forall x, y \in b, err(x, y) \leq \delta \tag{2}$$

The division into buckets can be done based on the representative value, the minimum tuple of each positive numerical attribute, as described in BAQ [3]. The size of the synopsis generated by BAQ is smaller than the original data sample, but there is still space for improvement in the bucket partitioning in order to obtain a smaller synopsis.

In this study, the average value p of each bucket b is considered as a representative value, and each bucket is defined in such a way that the following equation holds.

$$p = \frac{\sum_{x \in b} x}{|b|} \tag{3}$$

$$\forall x \in b, err(x, p) \leq \delta$$

We partition a given dataset into buckets with Algorithm 1, where Eq. (3) is used.

Algorithm 1: BucketGeneration

Data: Ascending sorted source data T, error threshold δ
Result: bucket $blk, mean, count$

```
 1 begin
 2 │   n := len(T)
 3 │   blk := T[0] // Create a new bucket.
 4 │   count := count + 1 // Number of elements in the blk
 5 │   mean := T[0] // Average value of blk
 6 │   for i := 1 → n - 1 do
 7 │   │   if is_possible_to_add(T[i], δ) then
 8 │   │   │   blk := blk ∪ T[i]
 9 │   │   │   count := count + 1
10 │   │   │   mean := (count * mean + T[i])/(count + 1) // Update the average
11 │   │   │   if i == n - 1 then
12 │   │   │   └   return blk, mean, count
13 │   │   else
14 │   │   │   return blk, mean, count
15 │   │   │   new_block(T[i])
16 │   │   │   count := 1
17 │   │   └   mean := T[i]
```

As the preprocessing step, let T be the data obtained by sorting tuples of the original data in ascending order based on their numerical attributes. The relative error threshold δ and the sorted data T as the input data. Next, initialize the data size n, *blk*, and the number of elements *count* and *mean* (lines 2–5). Then, in turn, generate buckets of tuples of the original data T based on Eq. (3) with Algorithm 2 (lines 7–14). Specifically, the division into buckets can be done based on the minimum positive and negative values of each numerical attribute.

Algorithm 2: IsPossibleToAdd(array, δ)

Data: Data set *array*, error threshold δ.
Result: *true* or *false*
1 **begin**
2 | *mean* = *array.mean*
3 | *min* = *array.min*
4 | **if** $|(mean - min)/min| \leq \delta$ **then**
5 | | **return** *true*
| **else**
6 | |_ **return** *false*

For example, when the minimum positive non-zero value x_1 is used as the criterion, the following equation can be used to generate buckets.

$$p_n \leq (1 + \delta)x_1 \tag{4}$$

Specifically, given numerical data $\{120, 130, 137, 140, 143, 146, 150, 155, 158,$ $161, 185, 190\}$ and a value of δ of 0.2, the first tuple 120 is used as the basis for group 1 (i.e., record 1 of the synopsis) in Eq. (4). Since the mean 144 up to the tenth tuple 161 is greater than $120*(1+0.2)$, and the mean 148 up to the eleventh tuple 185 is greater than $\{144, 120, 130, 137, 140, 143, 146, 150, 155, 158, 161\}$ is saved as group 1. Since 185 is the criterion for group 2 and Eq. (4) is satisfied up to 190, $\{185, 190\}$ are stored as group 2. So we get two groups, and we store the mean, minimum, maximum and number of elements of each group as buckets. For group 1, we finally get $\{144, 120, 161, 10\}$.

By the above method, the calculated buckets are larger than those derived from BAQ, which means the size of the generated synopsis is smaller than that of BAQ. Negative buckets are also created in the same way, and if there is a tuple with zero value, an additional bucket with only zero value is created.

3.2   Generating the Synopsis

In this study, we first summarize data and perform the join processing of correlated attributes based on selection conditions, and then generate a synopsis using the method proposed in [3].

Taking Query 1 as an example, when processing the complex queries, we need to process the join of four tables and the selection operations on multiple conditions. About Query 1, firstly we join the four tables and compute the bucket of (l_extendedprice, l_discount, c_acctbal, c_phone), generate the synopsis. Then, we join it with the synopsis for the remaining category attributes (c_custkey, c_name, n_name, c_address, o_orderdate), generate the last synopsis from the original table, and store it. When a query is given, it selects data from specific conditions based on the synopsis and performs an approximate calculation.

3.3 Query Processing

we describe the case where numerical attributes are used as selection conditions and category attributes are used in selection and the group-by clauses. The following **Query** 2 is an example.

```
SELECT l_orderkey,o_orderdate, COUNT(*)
FROM lineitem, orders
WHERE l_extendedprice > 11000 AND o_orderkey = l_orderkey
GROUP BY l_orderkey,o_orderdate
ORDER BY l_extendedprice
```

In response to this query, the schema of the corresponding synopsis by the proposed method is (orderkey, orderdate, price_min, price_max, SF), and an instance is created using the lineitem and orders tables. SF is the number of records within the range [price_min, price_max] for l_extendedprice. The synopsis does not include records for which $SF = 0$. In other words, in addition to the category attributes specified in the selection and grouping conditions, the buckets of numerical attributes specified in the selection conditions are considered as category attributes and grouped, and the number of tuples belonging to each group is used as records in the synopsis. The buckets are divided into three categories: 1) the whole bucket satisfies the condition, 2) the whole bucket does not satisfy the condition, and 3) a part of the bucket satisfies the condition.

In the proposed method, for the bucket in 1), the original query is rewritten as the following query to approximate the calculation.

```
SELECT l_orderkey,o_orderdate, sum(SF)
FROM synopsis
WHERE price_min > 11000
GROUP BY l_orderkey,o_orderdate
ORDER BY l_extendedprice
```

In the case of bucket 3), the following query computes it exactly.

```
SELECT l_orderkey,o_orderdate, count(*)
FROM lineitem, orders
WHERE l_extendedprice > 11000
  AND l_extendedprice < synopsis.price_max
GROUP BY l_orderkey,o_orderdate
ORDER BY l_extendedprice
```

When computing the COUNT function for the bucket in 3), the work will return the maximum value of the group that partially satisfies the condition price_max by synopsis, and return the number of records that satisfy the selection condition in the above query from the original data. Since the query is processed with respect to the original data, the error in the aggregation calculation in 3) is 0.

Since the error in computing the COUNT aggregation for the bucket in 2) is also zero, the proposed method can compute the COUNT function with zero error. we can guarantee the error δ for the SUM and AVG aggregate calculations based on our error work [11].

4   Experiment Results

4.1   Experiment Setup

Datasets. Two datasets were used in this study. The first one is Kaggle's temperature dataset with 7 columns (6 categorical columns and 1 numerical column). This dataset contains 36 years of temperature statistics for 321 cities. The second dataset is the well-known TPC-H dataset, where we used a 4-column (3 categorical columns and 1 numerical column) orders table in our experiments.

Queries. For a given query set, we generated 310 queries of each type in Sect. 5 for each dataset. For a given query set, for each dataset, we generated 310 queries of each type for each of the five clauses, including 2 MIN/MAX queries and 2 SUM/AVG queries without selection and condition, 50 COUNT queries, 50 MIN/MAX queries and 50 SUM/AVG queries with only category attribute as condition, 52 COUNT queries, 52 MIN/MAX queries and 52 SUM/AVG queries with numerical attributes in the selection condition. The above set of 310 queries can be used to generate an offline synopsis, and the size and generation time of the synopsis can be evaluated.

After that, to compare the processing time and relative error of the online queries, we generate the synopsis based on the attribute set of 310 queries. 2 MIN/MAX queries and 2 SUM/AVG queries without selection and conditions, 20 COUNT queries, 20 MIN/MAX queries and 20 SUM/AVG queries with only categorical conditions. 5 COUNT queries, 10 MIN/MAX queries, and 10 SUM/AVG queries with numerical attributes as selection conditions, and compared the results with BAQ. Since BAQ has been proven to be superior to the latest SAQP (Blink, Seek), Sketch, and Wavelet, we did not compare it with [3] and SAQP-based methods.

Metrics. We compare the proposed method with the BAQ from four angles. (1) the error between the result of estimation and the true value, (2) the size of the synopsis, and (3) the error between the result of estimation and the true value. (2) the size of the synopsis, (3) the time to process the query online (ms), and (3) the time to process the query online (ms), (4) the offline synopsis (4) the generation time of the offline synopsis (s). In the evaluation experiment, we run the proposed method and BAQ 20 times each and result in the average value.

4.2 Varying Error Bound

We set the error limits δ to 0.05, 0.1, 0.15, 0.2, and 0.25 and compare the proposed method with the BAQ.

Error. The results of the error calculation between the proposed method and the BAQ for the average value of the total query for the orders data, which has only positive numbers as numerical attributes, and the temperature data, which has both positive and negative numbers as numerical attributes, are shown in Fig. 1.

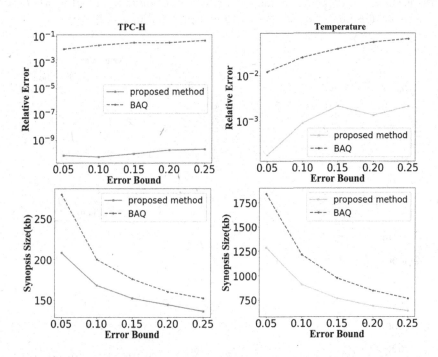

Fig. 1. The average relative error of 310 queries and the size of generated synopsis.

The relative error of the proposed method is much smaller: 10–20% of the BAQ for the proposed method when using the orders data, and 10–30% of the BAQ for the temperature data. For example, in the case of temperature data, in the error range of 0.05 to 0.25, the error of the proposed method is 0.002–0.01, while the error of the BAQ is 0.01–0.05. The reason is that the proposed method has 100% confidence level when the selection condition does not include any numerical attribute. As the error limit increases, the estimation error also increases, but since the limit δ corresponds to the worst estimation, the actual estimation error is always smaller than the specified error limit, and in fact is much smaller.

Size of Synopsis. The sizes of the synopsis generated by the proposed method and BAQ are shown in Fig. 1.

For the orders data, the synopsis size of the proposed method is 60–90% of that of the BAQ, while that of the temperature data is 50–80%.

Here is a simple analysis. The BAQ generates a synopsis based on the minimum value. The proposed method is obtained for each bucket by grouping on numerical attributes with the mean value as the representative value based on the positive and negative minimum values. The number of records in each bucket is larger for the proposed method than that for BAQ, and the size of the synopsis is smaller. Furthermore, increasing the error bound reduces the size of the synopsis because fewer buckets need to be generated to satisfy δ.

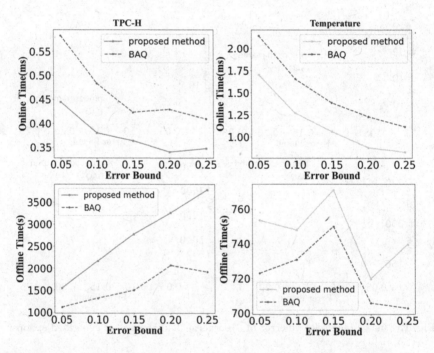

Fig. 2. The average time of the query processing and the average time of the synopsis generating.

Online Query Processing Time. For the orders and temperature data, we measured the time to process the query online. The results of the proposed method and the existing study BAQ are shown in Fig. 2.

Since the size of the synopsis of the proposed method was smaller than that of the BAQ, the proposed method was faster than the BAQ in answering the queries within 0.5 ms for orders and within 1.8 ms for temperature. As the error limit increased, the query processing time decreased. As the error limit increased, the synopsis also became smaller that the time to scan the synopsis became shorter.

Offline Time. The measured offline synopsis generation time for the orders and temperature datasets is shown in Fig. 2.

For the orders dataset, the proposed method took up to twice as long as BAQ. When δ is 0.25, the proposed method requires 3800 s, while the BAQ requires 2000 s. However, for temperature, the difference between the proposed method and the BAQ is smaller and both can be processed within 800 s.

The proposed method requires more time than the existing studies because it needs to calculate the average value of each bucket. Since the generation of the synopsis is performed offline, the impact on the user is considered to be small.

5   Conclusion

In this paper, we proposed an approach for approximate query processing that generates synopsis within the error bound and performs online query processing efficiently. We also implemented the approximate query processing framework with the above error guarantee and conducted evaluation experiments on two datasets. The experimental results show that the proposed approach outperforms the existing BAQ in terms of estimation error, online query processing time, and synopsis size, and can be efficiently applied to real-valued data within the error.

In the future, we plan to study approximate query processing methods for efficiently processing complex queries with more than one numerical attribute, and optimization of the summary synopsis.

Acknowledgment. This work was supported by KAKENHI(16H01722, 20K19804, and 21H03555).

References

1. Mozafari, B., Niu, N.: A handbook for building an approximate query engine. IEEE Data Eng. Bullet. **38**(3), 3–29 (2015)
2. Li, K., Li, G.: Approximate query processing: what is new and where to go? Data Sci. Eng. **3**, 379–397 (2018)
3. Li, K., Zhang, Y., Li, G., Tao, W., Yan, Y.: Bounded approximate query processing. IEEE TKDE **31**(12), 2262–2276 (2019)
4. Walenz, B., Sintos, S., Roy, S., Yang, J.: Learning to sample: counting with complex queries. PVLDB **13**(3), 389–401 (2019)
5. Cormode, G., Garofalakis, M., Haas, P.J., Jermine, C.: Synopsis for massive data: samples, histograms, wavelets, sketches. Found. Trends Databases **4**(1), 1–294 (2012)

6. Chaudhuri, S., Ding, B., Kandula, S.: Approximate query processing: no silver bullet. In: Proceedings SIGMOD, pp. 511–519. ACM (2017)
7. Agarwal, S.: Knowing when you're wrong: building fast and reliable approximate query processing systems. In: Proceedings SIGMOD, pp. 481–492. ACM (2014)
8. Ma, Q., Triantafillou, P.: DBEST: Revisiting approximate query processing engines with machine learning models. In: Proceedings SIGMOD, pp. 1553–1570. ACM (2019)
9. Pol, A., Jermaine, C.: Relational confidence bounds are easy with the bootstrap. In: Proceedings SIGMOD, pp. 587–598. ACM (2005)
10. Zeng, K., Gao, S., Gu, J., Mozafari, B., Zaniolo, C.: ABS: a system for scalable approximate queries with accuracy guarantees. In: Proceedings SIGMOD, pp. 1067–1070. ACM (2014)
11. Proof of Error Guarantee. https://www.db.is.i.nagoya-u.ac.jp/~ishikawa/docs/ErrorGuarantee.pdf. Accessed 28 Aug 2021

Author Index

Printed in the United States
by Baker & Taylor Publisher Services